Fractions, Ratios, and Percents

Contents

Introduction

Building a solid foundation in math is a student's key to success in school and in the future. This book will help students to develop the basic math skills that they will use every day. As students build on math skills that they already know and learn new math skills, they will see how much math connects to real life.

This book will help students to:

- develop math competence;
- acquire basic math skills and concepts;
- learn problem-solving strategies;
- apply these skills and strategies to everyday life;
- gain confidence in their own ability to succeed at learning.

Students who have self-confidence in their math skills often do better in other school areas, too. Mastering math helps students to become better learners and better students.

Ensure Student Success in Math

This book contains several features that help teachers to build the self-confidence of math students. This book enables the teacher to:

- reach students by providing a unique approach to math content;
- help students build basic foundational math skills;
- diagnose specific math intervention needs;
- provide individualized, differentiated instruction.

Assessment. An Assessment is included to serve as a diagnostic tool. The Assessment contains most of the math concepts presented in this book. An Assessment Evaluation Chart helps to pinpoint each student's strengths and weaknesses. Then, instruction can be focused on the math content each student needs. Each item in the Assessment is linked to a lesson in the book where students can hone their math skills.

Correlation to Standards. A Correlation to NCTM Standards is provided to allow teachers to tailor their teaching to standardized tests. This chart shows teachers at a glance which lessons cover the basic skills students are expected to master.

Lesson Format. Each lesson in the book is constructed to help students to master the specific concept covered in the lesson. A short introduction explains the concept. Then, a step-by-step process is used to work an example problem. Students are then given a short problem to work on their own. Finally, a page of practice problems that reinforce the concept is provided.

Glossary. Math has a language of its own, so a Glossary of math terms is included at the back of the book. Students can look up terms that confuse them, and they are directed to a specific page on which the term is explained or implemented.

Answer Key. A complete Answer Key is provided at the back of the book. The Answer Key includes the answers for the practice problems as well as explanations on how many of the answers are reached. These explanations can be useful to the teacher to explain why students might have answered incorrectly.

Graphic Organizers. Graphic organizers often help students to solve problems more easily. For that reason, a series of charts, detailed step-by-step processes, and various kinds of graphs and diagrams are supplied at the back of the book. Students can use these tools to help them solve the problems in the book or create their own problems.

Working Together to Help Students Achieve

No student wants to do poorly. There are many reasons students may be having problems with math. This book presents a well-organized, straightforward approach to helping students overcome the obstacles that may hold them back. This book and your instruction can help students to regain their footing and continue their climb to math achievement.

Correlation to NCTM Standards

Content Strands	Lesson
Number and Operations	
• work flexibly with fractions, decimals, and percents to solve problems	1, 3, 6, 8, 9, 10, 11, 12, 13, 14, 15, 16, 17, 18, 20, 21, 22, 23, 24, 25, 26, 27, 28, 29, 30, 31, 32, 33, 34, 35, 36, 37, 38, 39
• compare and order fractions, decimals, and percents efficiently and find their approximate locations on a number line	1, 2, 3, 4, 5, 6, 7, 8, 19, 26, 27, 28, 29, 30
• understand and use ratios and proportions to represent quantitative relationships	19, 20, 21, 22, 23, 24, 25
• understand the meaning and effects of arithmetic operations with fractions, decimals, and integers	3, 6, 8, 9, 10, 11, 12, 13, 14, 15, 16, 17, 18, 20, 21, 22, 23, 26, 27, 28, 29, 30, 31, 32, 33, 34, 35, 36, 37, 38, 39
• use the associative and commutative properties of addition and multiplication and the distributive property of multiplication over addition to simplify computations with integers, fractions, and decimals	15, 16
• select appropriate methods and tools for computing with fractions and decimals from among mental computation, estimation, calculators or computers, and paper and pencil, depending on the situation, and apply the selected methods	1, 2, 3, 4, 5, 6, 7, 8, 9, 10, 11, 12, 13, 14, 15, 16, 17, 18, 20, 21, 22, 23, 24, 25, 26, 27, 28, 29, 30, 31, 32, 33, 34, 35, 36, 37, 38, 39
• develop and analyze algorithms for computing with fractions, decimals, and integers and develop fluency in their use	3, 6, 9, 10, 11, 12, 13, 14, 15, 16, 17, 18, 20, 21, 22, 23, 24, 25, 26, 27, 28, 29, 30, 31, 32, 33, 34, 35, 36, 37, 38, 39
• develop, analyze, and explain methods for solving problems involving proportions, such as scaling and finding equivalent ratios	20, 21, 22, 23, 24, 25

Assessment

Write a fraction for each sentence.

1. Five out of 12 people in the club were absent. _____

2. There were 24 movies playing in the multiplex theater. I saw all 24. _____

3. Gretchen cut the pizza into 8 pieces and took 4 of them. _____

4. José chose 10 out of 10 of the markers for his poster. _____

Write each fraction in simplest form.

5. $\frac{6}{15}$ _____

6. $\frac{4}{6}$ _____

7. $\frac{6}{12}$ _____

8. $\frac{15}{18}$ _____

9. $\frac{6}{8}$ _____

10. $\frac{8}{16}$ _____

Find the least common denominator (LCD).

11. $\frac{1}{3}$ and $\frac{1}{6}$

12. $\frac{1}{2}$ and $\frac{5}{6}$

13. $\frac{1}{4}$ and $\frac{3}{7}$

14. $\frac{4}{5}$ and $\frac{1}{2}$

Add. Simplify if necessary.

15. Darren ate $\frac{1}{8}$ of a gallon of ice cream. His brother Darrell ate $\frac{3}{8}$ of the gallon. How much ice cream did they eat all together?

16. Lynn ate $\frac{1}{6}$ of the fruit salad. Her sister Leanne also ate $\frac{1}{6}$ of the fruit salad. How much of the fruit salad did they eat altogether?

17. $\frac{2}{5} + \frac{3}{10} =$

18. $3\frac{5}{8} + 2\frac{1}{2} =$

19. $\frac{1}{3} + \frac{1}{6} =$

20. $4\frac{5}{6} + 1\frac{1}{2} =$

Subtract. Simplify if necessary.

21. $\frac{5}{8} - \frac{3}{8} =$

22. $\frac{5}{6} - \frac{1}{6} =$

23. Aria had $\frac{1}{2}$ of a bag of apples. She gave $\frac{1}{6}$ of her apples away. What fraction of the bag of apples was left?

24. Calvin had $\frac{3}{4}$ of a bottle of orange juice. He drank $\frac{1}{6}$ of the juice. How much of the orange juice was left?

25. $6\frac{1}{3} - 2\frac{5}{6} =$

26. $5\frac{1}{8} - 2\frac{3}{4} =$

Multiply. Simplify if necessary.

27. How much is $\frac{3}{4}$ of $\frac{7}{8}$?

28. How much is $\frac{4}{5}$ of $\frac{5}{6}$?

29. $\frac{9}{10} \times \frac{5}{6} =$

30. $1\frac{1}{3} \times 1\frac{2}{3} =$

31. $3\frac{1}{8} \times 3\frac{1}{5} =$

32. $\frac{8}{9} \times \frac{15}{16} =$

Divide. Simplify if necessary.

33. $\frac{2}{3} \div \frac{1}{3} =$

34. How many times does $\frac{1}{4}$ go into $\frac{5}{8}$?

35. $2\frac{1}{2} \div 1\frac{1}{2} =$

36. $2\frac{4}{5} \div 1\frac{1}{2} =$

37. $\frac{4}{5} \div \frac{1}{5} =$

38. $4\frac{1}{5} \div 2\frac{1}{3} =$

Use the figures below to answer numbers 39 and 40. Be sure your answer is in simplest form.

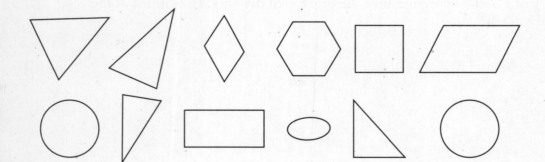

39. What is the ratio of circles to triangles?

40. What is the ratio of triangles to total figures?

Write each rate in simplest form.

41. 12 pages in 3 minutes

42. 80 kilometers in 4 hours

43. 120 words in 2 minutes

44. 60 miles on 3 gallons of gas

Solve.

45. Roy bought 6 cans of juice for $1.32. What is the price per can?

46. A 3-pound melon sells for $2.10. What is the price per pound?

Write = or ≠ in the box.

47. $\dfrac{3}{8}$ ☐ $\dfrac{6}{10}$

48. $\dfrac{4}{9}$ ☐ $\dfrac{12}{27}$

49. $\dfrac{5}{8}$ ☐ $\dfrac{16}{10}$

50. $\dfrac{4}{7}$ ☐ $\dfrac{12}{21}$

Solve for *n*.

51. $\dfrac{6}{n} = \dfrac{9}{6}$

52. $\dfrac{n}{20} = \dfrac{45}{15}$

53. $\dfrac{4}{n} = \dfrac{12}{3}$

54. $\dfrac{n}{8} = \dfrac{48}{16}$

Solve.

55. Two out of every 9 students were accepted into the marching band. If 225 students applied, how many were accepted?

56. Ramon is paid by the hour. During one 35-hour week he earned $420. What is Ramon's hourly pay?

Write each decimal or fraction as a percent.

57. 0.065

58. $\dfrac{7}{8}$

59. $\dfrac{5}{16}$

60. 0.062

Write each percent as a fraction in simplest form.

61. 32%

62. 125%

63. 42%

64. 115%

Solve.

65. Find 22% of 60.

66. What percent of 40 is 24?

67. 1.2 is 40% percent of what number?

68. 22 is what percent of 88?

69. Meg sold 75% of her collection of DVDs. She sold 30 DVDs. How many were in her collection?

70. At *Gene's Jeans Store,* all items go on sale for 15% off if they have not sold after being in the store for one week. If a pair of jeans is originally priced at $40, how much do they cost 10 days later?

Find the interest.

71. $600 deposited for 2 years at annual rate of 4% simple interest

72. $4,000 deposited for 9 months at 5.5% simple interest

Find the balance in the account.

73. Principal is $250; compound interest rate is 5% compounded semi-annually; time is 1 year

74. Principal is $100; compound interest rate is 7.5% compounded annually; time is 2 years

Assessment Evaluation Chart

Note the number of each assessment item that you missed. Then use the Lesson Review list to find more practice problems.

ITEM NUMBERS	SKILL	LESSONS FOR REVIEW
1–4	Understanding Fractions	Lessons 1–3
5–10	Writing Mixed Numbers in Simplest Form	Lessons 5–7
11–14	Finding the Least Common Denominator (LCD)	Lesson 4
15–16	Adding Fractions with Like Denominators	Lesson 9
17–20	Adding Fractions with Unlike Denominators	Lessons 10–11
18, 20	Adding Mixed Numbers	Lesson 11
21–22	Subtracting Fractions with Like Denominators	Lesson 12
23–26	Subtracting Fractions with Unlike Denominators	Lessons 13–14
25–26	Subtracting Mixed Numbers	Lesson 14
27–32	Multiplying Fractions	Lessons 15–16
33–38	Dividing Fractions	Lessons 17–18
39–40	Ratios	Lesson 19
41–44	Rates	Lesson 20
45–46	Learning About Unit Rates	Lesson 21
47–50	Proportions	Lessons 22–23
51–56	Problem Solving Using Proportions	Lessons 24–25
57–64	Converting Fractions, Decimals, and Percents	Lessons 26–30
65–70	Creating Percent Equations from Word Problems	Lessons 34–35
71–72	Simple Interest	Lesson 38
73–74	Compound Interest	Lesson 39

Concentrate on any skill section in which you missed one or more problems.

LESSON 1   Understanding Fractions

A **fraction** is a way of comparing part of a group to the whole group.

Three of the 7 number cards show the number 1. You can also say that $\frac{3}{7}$ of the cards in the group of cards show the number 1.

In the fraction $\frac{3}{7}$, the number 3 shows part of the group. This number is called the **numerator**.

The total number of cards is called the **denominator**.

$$\frac{\text{part}}{\text{whole}} = \frac{\text{numerator}}{\text{denominator}} = \frac{3}{7}$$

Example

A pizza is cut into 8 equal slices. You and your friends eat 5 slices. Write a fraction to show the part of the pizza that is left.

STEP 1 Identify the denominator.
The total number of slices is the whole. The pizza was cut into a total of 8 slices, so the denominator is 8.

$\frac{}{8}$

STEP 2 Identify the numerator.
The part of the whole is the numerator. There are 3 pieces of pizza left, so the numerator is 3.

$\frac{3}{8}$

STEP 3 Write the fraction.
$$\frac{\text{part}}{\text{whole}} = \frac{\text{numerator}}{\text{denominator}} = \frac{3}{8}$$
$\frac{3}{8}$ of the pizza is left.

ON YOUR OWN

Three of the last 5 songs on the radio were hits from last year. Write a fraction to show how many of the songs were hits from last year.

Practice

Building Skills

Write a fraction.

1. What part of a dollar is shown?

2. Of all the fruit shown, what fraction is apples?

3. What fraction of a gallon is shown? (4 quarts = 1 gallon)

Write a fraction.

4. Out of 18 friends at the picnic, 11 live on the same street. What part of the friends live on the same street?

5. A dozen is equal to 12. What part of a dozen is 11?

6. There are 12 cats living at the farm. Of these, 7 are female. What part of the cats are female?

7. If a day is equal to 24 hours, what part of a day is 8 hours?

8. If a foot is equal to 12 inches, what part of a foot is 7 inches?

9. There are 24 students in the class. Nineteen of the 24 students are right-handed. What part of the students are right-handed?

Problem Solving

Solve.

10. Jessica works 40 hours per week. So far this week, she has worked 9 hours. What part of her workweek has she completed?

11. One hour is equal to 60 minutes. What part of an hour is 40 minutes?

12. Tanya can fit 16 pictures on each page of her photo album. If she has 12 pictures, what fraction of a page will she need to use?

13. Devon hit 57 free throws in 60 tries. What fraction of free throws did he complete?

LESSON ② Equivalent Fractions

Sometimes 2 fractions can have the same value even when they do not look the same. Look at the 2 pizzas. They are the same size. One pizza has 8 slices. The other has 4 slices. The pizzas have different numbers of slices, but the amount of the whole pizza is the same. The 2 pizzas are equal or **equivalent.** Now look at the part of each pizza covered with onions.

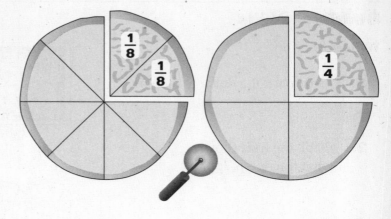

We can use an equivalent fraction to describe the pizza slices with onions.

$$\frac{2 \text{ slices}}{8 \text{ slices}} = \frac{1 \text{ slice}}{4 \text{ slices}} \text{ or}$$

$$\frac{2}{8} = \frac{1}{4}$$

Equivalent fractions name the same amount. You make fractions equivalent by multiplying or dividing the numerator and denominator by the same number.

Example

Find an equivalent fraction for $\frac{6}{16}$.

STEP 1 Choose a number that you can multiply or divide into both the numerator and denominator.
You can multiply both 6 and 16 by 2.

STEP 2 Multiply or divide the numerator and denominator.
Multiply the numerator and denominator by the same number, 2.

$$\frac{6}{16} \times \frac{2}{2} = \frac{12}{32}$$

STEP 3 Write the equivalent fractions.
$\frac{12}{32}$ is equivalent to $\frac{6}{16}$.

$$\frac{6}{16} = \frac{12}{32}$$

ON YOUR OWN

Find an equivalent fraction for $\frac{10}{12}$.

Practice

Building Skills

Find an equivalent fraction.

1. $\dfrac{4}{8}$ **2.** $\dfrac{10}{15}$ **3.** $\dfrac{6}{12}$ **4.** $\dfrac{18}{24}$

Circle the equivalent fraction.

5. $\dfrac{1}{2} = \dfrac{3}{10} \quad \dfrac{6}{12} \quad \dfrac{4}{7} \quad \dfrac{3}{9} \quad \dfrac{2}{8}$

6. $\dfrac{3}{5} = \dfrac{3}{10} \quad \dfrac{7}{12} \quad \dfrac{6}{10} \quad \dfrac{6}{20} \quad \dfrac{10}{15}$

7. $\dfrac{2}{3} = \dfrac{8}{10} \quad \dfrac{6}{9} \quad \dfrac{4}{12} \quad \dfrac{3}{9} \quad \dfrac{2}{24}$

8. $\dfrac{9}{10} = \dfrac{3}{4} \quad \dfrac{18}{21} \quad \dfrac{4}{5} \quad \dfrac{3}{9} \quad \dfrac{27}{30}$

Problem Solving

Find the equivalent fraction.

9. Manny has read 6 of the 10 books in a popular mystery series. How many fifths is this?

10. Anna has converted $\frac{5}{8}$ of her old videocassettes to DVDs. How many sixteenths is this?

11. Vinny shelved 30 of the 40 books that were returned to the library today. How many fourths is this?

12. Raquel invited 20 people to her party, and 16 of them showed up. Write an equivalent fraction to describe how many of the invited people came to the party.

13. You have collected 25 of the 50 state quarters. Write an equivalent fraction to show this number.

14. You did 25 of the 30 push-ups required in gym class. Find an equivalent fraction when 5 is the numerator.

LESSON ③ Least Common Multiple

Suppose you do volunteer work at a local food pantry every 5 days. Your friend volunteers there every 4 days. How often will the two of you work together in the first month?

First compare the multiples of 5 and 4. Circle the multiples they have in common. Then find the lowest multiple that 4 and 5 have in common.

> 5: 5, 10, 15, 20, 25, 30, 35, 40, 45, 50
> 4: 4, 8, 12, 16, 20, 24, 28, 32, 36, 40

The **least common multiple** (LCM) of 4 and 5 is 20.

If you get to work together every 20 days, then in 1 month you will work together on 1 day.

Example

Matt plays the piano every 3 days. Lola plays the piano every 4 days. If they both played today, on what day will they both play again?

STEP 1 List several multiples of the larger number.
4: 4, 8, 12, 16, 20, 24, 28, 32

STEP 2 List several multiples of the smaller number.
3: 3, 6, 9, 12, 15, 18, 21, 24

STEP 3 Find the multiple or multiples that are common to both numbers.
4: 4, 8, 12, 16, 20, 24, 28, 32
3: 3, 6, 9, 12, 15, 18, 21, 24

The multiples common to both 3 and 4 are 12 and 24.

STEP 4 Compare the multiples. The multiple that is lower in value is the least common multiple.
12 < 24

Twelve is the least multiple they have in common. So the LCM of 3 and 4 is 12. Matt and Lola will both play the piano on the twelfth day.

ON YOUR OWN

Rachel swims 10 laps every fourth day and uses free weights every fifth day. If she did both today, in how many days will she once again swim 10 laps and use free weights on the same day?

Practice

Building Skills

Find the least common multiple (LCM) for each pair of numbers.

1. 9 and 12

2. 4 and 8

3. 6 and 9

4. 8 and 9

5. 6 and 8

6. 10 and 15

7. 6 and 7

8. 5 and 8

9. 4 and 10

Problem Solving

Use the LCM to solve these problems.

10. Jared finishes a book in 6 days. Marta finishes a book in 4 days. If they both start reading on the same day, how many days will it take before they both finish a book on the same day?

11. Juan receives a news e-mail every 7 days and a music e-mail every 5 days. He received both of them today. In how many more days will he again receive both e-mails on the same day?

12. During rush hour, the express bus arrives every 8 minutes, while the regular bus comes every 10 minutes. If both buses just went by, how many more minutes will it take for both of them to again show up at the same time?

13. Nakita goes to swim practice every second day. Emily goes to swim practice every seventh day. If they both went to practice today, in how many days will they both go to practice again on the same day?

14. Your training program requires you to run every second day and to do weight training every third day. If you did both today, what is the next day on which you will have to run and do weight training?

15. Damon does dark laundry every 3 days and light laundry every 5 days. How often does he have to do both loads on the same day?

LESSON (4) Comparing and Ordering Fractions

As part of your lunch, you have 2 granola bars. You decide to share with your friends. One friend gets $\frac{2}{3}$ of 1 bar. The other friend gets $\frac{3}{4}$ of the other bar. Look at the granola bars. You can see that $\frac{3}{4} > \frac{2}{3}$.

To compare fractions, the fractions must have the same denominator. You can use your knowledge of equivalent fractions to rewrite each fraction so that it has the same denominator. To find the common denominator, use the least common multiple (LCM) of the denominators. This is also called the **least common denominator** (LCD). When the denominators are the same, compare the numerators to see which fraction is greater.

$$\frac{\text{numerator}}{\text{denominator}} = \frac{3}{4}$$

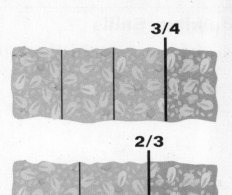

3/4

2/3

Example

Compare. Use <, >, or = signs.
$\frac{3}{7}$ $\frac{2}{5}$

STEP 1 Find the least common denominator (LCD).
List the multiples of 5 and 7.
5: 5, 10, 15, 20, 25, 30, 35, 40, . . .
7: 7, 14, 21, 28, 35, 42, 49, 56, . . .

The least common multiple (LCM) of the denominators is 35. So the LCD is 35.

STEP 2 Write equivalent fractions using the LCD as the new denominator.

$$\frac{3}{7} = \frac{}{35} \qquad \frac{2}{5} = \frac{}{35}$$

STEP 3 Divide the LCD by the denominators.
$35 \div 7 = 5 \qquad 35 \div 5 = 7$

STEP 4 Multiply the numerator and denominator of each fraction by the answers in Step 3.

$$\frac{3}{7} \times \frac{5}{5} = \frac{15}{35} \qquad \frac{2}{5} \times \frac{7}{7} = \frac{14}{35}$$

STEP 5 Compare the numerators.
$\frac{3}{7} > \frac{2}{5}$

$15 > 14$, so $\frac{15}{35} > \frac{14}{35}$

(ON YOUR OWN)

Kirsten has completed $\frac{1}{3}$ of her homework. Brendan has completed $\frac{2}{5}$ of his homework. Use <, >, or = to show who has completed the greater part of the homework assignment.

Practice

Find the LCD and then compare the numerators.

Building Skills

Compare. Use >, <, or = signs.

1. $\dfrac{5}{6}$ ☐ $\dfrac{1}{4}$

2. $\dfrac{1}{5}$ ☐ $\dfrac{2}{10}$

3. $\dfrac{3}{4}$ ☐ $\dfrac{7}{8}$

4. $\dfrac{4}{5}$ ☐ $\dfrac{2}{3}$

5. $\dfrac{2}{5}$ ☐ $\dfrac{1}{2}$

6. $\dfrac{3}{8}$ ☐ $\dfrac{1}{4}$

7. $\dfrac{9}{10}$ ☐ $\dfrac{3}{4}$

8. $\dfrac{3}{6}$ ☐ $\dfrac{4}{12}$

9. $\dfrac{5}{8}$ ☐ $\dfrac{3}{7}$

Write these fractions in order from least to greatest.

10. $\dfrac{2}{5}$, $\dfrac{7}{10}$, $\dfrac{1}{2}$

11. $\dfrac{2}{3}$, $\dfrac{5}{6}$, $\dfrac{1}{2}$

Problem Solving

Solve each problem.

12. Lamont stayed at the game for $\frac{1}{2}$ hour. Jerry was there for $\frac{3}{4}$ hour. Who stayed at the game longer?

13. Christine rode her bike $\frac{4}{5}$ of a mile. Barbara ran $\frac{9}{10}$ of a mile. Gary jogged $\frac{1}{2}$ mile. Who traveled the farthest?

14. One type of juice provides $\frac{2}{3}$ of a person's daily requirement of vitamin C. Another provides $\frac{3}{4}$ of the daily requirement. Which juice is less nutritious?

15. Suppose you have to organize wrenches by size. Place these three wrenches in order from smallest to largest: $\frac{1}{4}$ inch, $\frac{3}{16}$ inch, $\frac{1}{8}$ inch.

LESSON 5 — Mixed Numbers as Improper Fractions

A mixed number is a number written as a whole number and a fraction. An **improper fraction** is a fraction whose numerator is *equal to or greater than* its denominator. For example, $\frac{11}{10}$, $\frac{3}{2}$, $\frac{7}{4}$, and $\frac{20}{20}$ are improper fractions.

An improper fraction can be written as a mixed number. To change a mixed number into an improper fraction, multiply the whole number by the denominator and then add the numerator. The answer is your new numerator. You keep the original denominator.

denominator $\times$ whole number + numerator = new numerator

or

$$3\frac{1}{2} \rightarrow \frac{(2 \times 3) + 1}{2} = \frac{7}{2}$$

whole number $\rightarrow 3\frac{1}{2} \begin{array}{l} \leftarrow \text{numerator} \\ \leftarrow \text{denominator} \end{array}$

Example

Write $4\frac{1}{3}$ as an improper fraction.

STEP 1 Multiply the denominator by the whole number.
$3 \times 4 = 12$

STEP 2 Add the numerator.
$12 + 1 = 13$

STEP 3 Write that total as the numerator, keeping the denominator the same.
$4\frac{1}{3}$ written as an improper fraction is $\frac{13}{3}$.

$\boxed{\dfrac{13}{3}}$

ON YOUR OWN

Write $3\frac{5}{8}$ as an improper fraction.

Practice

> The denominator of the mixed number is the denominator of the improper fraction.

Building Skills

Write each mixed number as an improper fraction.

1. $2\frac{5}{8} =$

2. $3\frac{4}{5} =$

3. $4\frac{5}{6} =$

4. $2\frac{9}{10} =$

5. $6\frac{2}{3} =$

6. $4\frac{1}{2} =$

7. $3\frac{2}{5} =$

8. $4\frac{1}{8} =$

9. $1\frac{7}{8} =$

Problem Solving

Write each mixed number as an improper fraction.

10. Milagro needs to add $2\frac{1}{4}$ quarts of oil to her car's engine. Change $2\frac{1}{4}$ to an improper fraction.

11. The baseball game was rained out after $4\frac{2}{3}$ innings. Change $4\frac{2}{3}$ to an improper fraction.

12. Lily bought $1\frac{5}{8}$ pounds of egg salad. How many eighths of egg salad did she buy?

13. Emile's band rehearsed for $1\frac{1}{2}$ hours. How many half-hours was that?

14. Julia ran $2\frac{1}{4}$ miles around the track. How many $\frac{1}{4}$-mile laps did she run?

15. George has $2\frac{3}{4}$ yards of electrical wire. He wants to cut the wire into $\frac{1}{4}$-yard lengths. How many pieces of wire will he get?

LESSON 6 Renaming Improper Fractions

Just as you can **rename**, or change, a mixed number to an improper fraction, you can change an improper fraction to a mixed number. When you see an improper fraction, divide the numerator by the denominator. The result is the whole number part of your mixed number. Any remainder is the numerator of the proper fraction.

You can think of the fraction bar as a division sign.

$\frac{19}{4}$ means *nineteen divided by four*.

Example

Change $\frac{19}{4}$ into a mixed number.

$19 \rightarrow$	numerator
$4 \rightarrow$	denominator

STEP 1 Divide the numerator by the denominator. The first step of dividing gives you the whole number part of the mixed number.

$$\begin{array}{r} 4 \\ 4\overline{)19} \\ -16 \\ \hline 3R \end{array}$$

STEP 2 The remainder then becomes the numerator of the fraction.

The remainder of 3 is the numerator in $4\frac{3}{4}$.

STEP 3 Simplify the fraction if necessary.

The fraction $\frac{3}{4}$ is in simplest form.

Therefore, $\frac{19}{4} = 4\frac{3}{4}$.

ON YOUR OWN

Change $\frac{16}{10}$ to a mixed number. Simplify if possible.

Practice

Building Skills

Rename each improper fraction.

1. $\dfrac{17}{10}$ 2. $\dfrac{23}{5}$ 3. $\dfrac{18}{6}$

4. $\dfrac{21}{4}$ 5. $\dfrac{45}{8}$ 6. $\dfrac{37}{7}$

7. $\dfrac{29}{6}$ 8. $\dfrac{61}{8}$ 9. $\dfrac{47}{5}$

Problem Solving

State the number in each problem as an improper fraction and then as a mixed number or whole number.

10. Linda said she could put a toy together in 90 seconds. What part of a minute is this? (1 minute = 60 seconds)

11. Jake walked for $\frac{7}{4}$ of an hour for exercise. Rewrite this improper fraction.

12. Vilma made 5 quarts of chili for the family reunion. How many gallons is this? (4 quarts = 1 gallon)

13. How many pairs of socks can be made from a laundry pile of 27 socks?

14. The student council is selling note cards in sets of 9. How would 64 cards be packaged? Write the answer as a mixed number.

15. Zach has to cut wooden molding into 9 pieces that are each half a yard long. What is the total length of the molding in yards written as a mixed number?

LESSON 7

Writing Mixed Numbers in Simplest Form

When you look at a mixed number, pay special attention to the fraction. Look for 2 things:

- Is the fraction proper or improper?
- Is the fraction in simplest form?

If the fraction is improper:

- Divide the numerator by the denominator.
- Add the whole number part of the answer to the existing whole number.
- Use the remainder as the numerator of the fraction.
- Keep the denominator the same as when you started.

If the fraction is a proper fraction:

- Find a number that divides evenly into the numerator and denominator.
- Divide the numerator and denominator by this number.
- Rewrite the mixed number including the simplified fraction.

Example

Simplify $5\frac{10}{8}$.

STEP 1 Is the fraction proper or improper?
The fraction is improper because the numerator is larger than the denominator.

STEP 2 Write the improper fraction as a mixed number.
$10 \div 8 = 1R2 \rightarrow 1\frac{2}{8}$

STEP 3 Add the renamed improper fraction to the whole number you started with.

$$5 + 1\frac{2}{8} = 6\frac{2}{8}$$

STEP 4 Simplify the fraction if necessary.
The fraction $\frac{2}{8}$ is not in simplest terms. To simplify, divide both the numerator and denominator by a number that divides evenly into both, 2.

$$\frac{2}{8} \div \frac{2}{2} = \frac{1}{4}, \text{ so } 5\frac{10}{8} = 6\frac{1}{4}.$$

$5\frac{10}{8}$ is equivalent to $6\frac{1}{4}$.

ON YOUR OWN

Neil has two pieces of speaker cable with a combined length of $2\frac{15}{12}$ yards. Simplify the mixed number.

Practice

Building Skills

Rename each mixed number in simplest form.

1. $3\frac{12}{10}$

2. $2\frac{6}{8}$

3. $6\frac{4}{6}$

4. $5\frac{9}{6}$

5. $2\frac{5}{10}$

6. $4\frac{12}{16}$

7. $7\frac{9}{12}$

8. $3\frac{10}{8}$

9. $5\frac{8}{12}$

Problem Solving

Write each mixed number in simplest form.

10. After a week of pet sitting, Bik has $2\frac{4}{8}$ cans of dog food left. Write this mixed number in simplest form.

11. The cafeteria used $8\frac{12}{10}$ gallons of pancake batter for the charity breakfast. What is this number in simplest form?

12. Amanda ran the race in $30\frac{6}{10}$ seconds. Write this number in simplest form.

13. The users of the computer room went through $7\frac{20}{25}$ packages of printer paper last week. What is this number in simplest form?

14. Wen-wa has $5\frac{12}{16}$ yards of material. Write this number in simplest form.

15. Amy says she has $7\frac{12}{10}$ boxes of comic books. How many boxes does she really have?

LESSON 8 Fractions and Decimals

A fraction is a way to show division. To change a fraction to a decimal, divide the numerator by the denominator until the answer ends evenly or the final digit(s) repeat. Remember to put a decimal point followed by 2 zeros after the digit of the numerator to set up the division.

$$\frac{3}{4} = 3 \div 4; \quad 4\overline{)3.00}^{\,0.75}$$

denominator $\overline{)}$ numerator

Example 1

Change $\frac{7}{8}$ to a decimal.

STEP 1 Divide the numerator by the denominator.
Place a decimal point after the dividend (the amount being divided).

STEP 2 Place a decimal point in the quotient (the answer).
Place the decimal point directly above the decimal point in the dividend.

$$\frac{7}{8} = 0.875$$

$$\begin{array}{r} .875 \\ 8\overline{)7.000} \\ -64 \\ \hline 60 \\ -56 \\ \hline 40 \\ -40 \\ \hline 0 \end{array}$$

To write a decimal as a fraction, write the decimal number as the numerator but do not include the decimal point. In the denominator, write a 1 followed by as many zeros as the number of digits behind the decimal point. Simplify the fraction if necessary.

Example 2

Change 0.225 to a fraction.

STEP 1 Write the number with no decimal point as the numerator.
225

STEP 2 Write a 1 and add zeros.
Count the number of places after the decimal point in the original number. In this problem, there are 3 places. You would add 3 zeros after the 1.

$$\frac{225}{1000}$$

STEP 3 Simplify the fraction as necessary.

$$0.225 = \frac{9}{40}$$

$$\frac{225}{1000} \div \frac{25}{25} = \frac{9}{40}$$

(**ON YOUR OWN**)

Change $\frac{5}{8}$ to a decimal.

Practice

Building Skills

Write each decimal as a fraction and each fraction as a decimal.

1. $\dfrac{1}{2} =$

2. $\dfrac{3}{8} =$

3. $\dfrac{4}{5} =$

4. $0.95 =$

5. $0.125 =$

6. $0.72 =$

7. $\dfrac{2}{5} =$

8. $\dfrac{7}{25} =$

9. $0.625 =$

Problem Solving

Solve each problem by changing the fraction to a decimal or the decimal to a fraction.

10. Maria has 3 quarters. What part of a dollar is 3 quarters? Write this as a decimal.

11. Bret only had time to complete $\frac{7}{8}$ of his run today. Write this as a decimal.

12. The students bought the books for $\frac{1}{4}$ of the original cost. Show the discount as a decimal.

13. Elian ate $\frac{5}{8}$ of his pizza. How do you write that as a decimal?

14. Jennie spent $0.85 on a juice drink. Write this as a fraction.

15. Phil finished 0.625 of his homework. Write the decimal as a fraction.

LESSON 9 Adding with Like Denominators

Before you can add fractions, they must have the same, or common, denominators. If the fractions already have the same denominator, just add the numerators.

$$\frac{\text{numerator}}{\text{denominator}} + \frac{\text{numerator}}{\text{denominator}} = \frac{\text{numerator} + \text{numerator}}{\text{denominator}}$$

$$\frac{2}{6} + \frac{3}{6} = \frac{5}{6}$$

When you add fractions, you add the numerators but you <u>do not</u> add the denominators.

After you add, you must check to make sure that the fraction is in simplest form.

Example

Add. $\frac{2}{5} + \frac{1}{5}$

STEP 1 If both fractions have the same denominator, add the numerators.

$$\frac{2 + 1}{5} = \frac{3}{5}$$

STEP 2 If the answer is an improper fraction, rename it as a mixed number.
In this case, $\frac{3}{5}$ is a proper fraction, so there is no need to rename the fraction.

STEP 3 Simplify the fraction.
The fraction $\frac{3}{5}$ is already in simplest form.

$$\frac{2}{5} + \frac{1}{5} = \frac{3}{5}$$

ON YOUR OWN

Julio practiced his trumpet $\frac{3}{4}$ hours yesterday and $\frac{3}{4}$ hours today. How long did he practice in all? Simplify if possible.

Practice

Building Skills

Fractions that are answers must be in their simplest form.

Add. Simplify if possible.

1. $\dfrac{1}{8} + \dfrac{5}{8} =$

2. $\dfrac{1}{7} + \dfrac{1}{7} =$

3. $\dfrac{3}{20} + \dfrac{7}{20} =$

4. $\dfrac{1}{4} + \dfrac{1}{4} =$

5. $\dfrac{1}{10} + \dfrac{7}{10} =$

6. $\dfrac{4}{15} + \dfrac{8}{15} =$

7. $\dfrac{7}{12} + \dfrac{11}{12} =$

8. $\dfrac{4}{9} + \dfrac{5}{9} =$

9. $\dfrac{13}{16} + \dfrac{7}{16} =$

Problem Solving

Solve.

10. At the community center, Maureen answered $\frac{1}{5}$ of the phone calls. Donna answered $\frac{3}{5}$ of the phone calls. What part of the phone calls did they answer in all?

11. Benny read $\frac{2}{5}$ of his book last night. He read another $\frac{2}{5}$ this morning. What part of the book has he read so far?

12. Jessica gave $\frac{3}{6}$ of last year's school clothes to charity and $\frac{1}{6}$ to her cousin. What part of her clothing did she give away?

13. If $\frac{3}{8}$ of a class is in the school chorus and $\frac{1}{8}$ is in the school band, what part of the class is involved in either chorus or band?

14. In history class, $\frac{3}{5}$ of your grade is based on test scores and $\frac{1}{5}$ on class participation. How much of your total grade is this?

15. The treasurer reports that $\frac{3}{10}$ of your class paid their class dues at the start of the school year. Another $\frac{2}{10}$ paid dues after the winter break. What part of the students paid their class dues?

LESSON ⑩ Adding with Unlike Denominators

When you add fractions, their denominators must be the same. But what happens when you need to add fractions that have different denominators? You have to rename the fractions as equivalent fractions so that all fractions have the same denominator. Then you add the numerators. Remember, equivalent fractions are fractions that name the same value.

Example

Add. $\frac{1}{4} + \frac{1}{3}$

STEP 1 Find the least common denominator (LCD).
4: 4, 8, **12**, 16, 20
3: 3, 6, 9, **12**, 15
The LCD of 3 and 4 is 12.

STEP 2 Write equivalent fractions with the LCD. Remember that to get the equivalent fraction you must multiply the numerator and denominator by the same number.

$$\frac{1}{4} \times \frac{3}{3} = \frac{3}{12}$$
$$\frac{1}{3} \times \frac{4}{4} = \frac{4}{12}$$

STEP 3 Add the numerators.

$$\frac{3+4}{12} = \frac{7}{12}$$

STEP 4 If necessary, rewrite as a mixed number and/or simplify the answer.
In this case, neither step is necessary because $\frac{7}{12}$ is both a proper fraction and in simplest form.
$$\frac{1}{4} + \frac{1}{3} = \frac{7}{12}$$

ON YOUR OWN

Ramon jogged for $\frac{2}{3}$ hour. Crystal jogged for $\frac{1}{2}$ hour. How long did they both jog?

Practice

Building Skills

Add.

1. $\dfrac{1}{4} + \dfrac{2}{3} =$

2. $\dfrac{1}{2} + \dfrac{3}{5} =$

3. $\dfrac{3}{8} + \dfrac{1}{6} =$

4. $\dfrac{1}{9} + \dfrac{5}{6} =$

5. $\dfrac{2}{3} + \dfrac{5}{6} =$

6. $\dfrac{4}{5} + \dfrac{1}{2} =$

7. $\dfrac{7}{8} + \dfrac{5}{6} =$

8. $\dfrac{3}{4} + \dfrac{4}{5} =$

9. $\dfrac{9}{10} + \dfrac{1}{4} =$

Problem Solving

Solve.

10. Alvaro practiced for $\frac{1}{2}$ hour yesterday and $\frac{1}{4}$ hour today. How long did he practice altogether?

11. Pam worked on her car for $\frac{3}{4}$ hour on Saturday and for $\frac{5}{6}$ hour on Sunday. How much time did she spend on her car over the weekend?

12. Drayton's video project filled $\frac{2}{8}$ of a recordable CD. Rob's video filled $\frac{1}{3}$ of a disc. How much disc space did both projects use?

13. Ann spent $\frac{2}{3}$ hour fixing her bike and then $\frac{5}{6}$ hour riding the bike. How many hours did she spend altogether?

14. Terry spent $\frac{1}{3}$ of his paycheck on a haircut and $\frac{2}{7}$ on school lunches for the week. What portion of his earnings did he spend that week?

15. Yeng spent $\frac{2}{8}$ of her day updating her web site and $\frac{3}{6}$ of that day at her part-time job. How much of the day did she spend on those two activities?

LESSON 11 Adding Mixed Numbers

On a typical afternoon at your job, you work for $2\frac{1}{2}$ hours, you take a break, and then you work for another $1\frac{3}{4}$ hours. If you wanted to find the total number of hours you worked, you would add mixed numbers.

When you add mixed numbers, you first add the whole numbers. Then you add the fractions. Remember that when you add any fractions, the denominators must be the same.

Example

Add. $2\frac{3}{8} + 3\frac{1}{4}$

STEP 1 Find the least common denominator (LCD) of the fractions.
List the multiples of 4 and 8.
4: 4, 8, 12, 16
8: 8, 16, 24, 32
The LCD of 4 and 8 is 8.

STEP 2 Write equivalent fractions using the LCD.

$$\frac{1}{4} \times \frac{2}{2} = \frac{2}{8}$$

STEP 3 Add the fractions.

$$\frac{2}{8} + \frac{3}{8} = \frac{5}{8}$$

STEP 4 Add the whole numbers.
$2 + 3 = 5$

STEP 5 Combine the whole number part and the fraction, and simplify the fraction if necessary.
The fraction $5\frac{5}{8}$ is in simplest form.

$$2\frac{3}{8} + 3\frac{1}{4} = 5\frac{5}{8}$$

ON YOUR OWN

Sara has collected $3\frac{3}{4}$ pounds of newspaper for recycling. Henry has collected $4\frac{1}{2}$ pounds. How much do they have together?

Practice

Building Skills

The denominators of the fractions must be the same before you add.

Add.

1. $2\frac{1}{6} + 2\frac{2}{3} =$

2. $5\frac{2}{5} + 2\frac{3}{10} =$

3. $1\frac{1}{3} + 2\frac{1}{4} =$

4. $3\frac{5}{8} + 2\frac{1}{6} =$

5. $3\frac{1}{2} + 4\frac{2}{5} =$

6. $6\frac{2}{3} + 2\frac{3}{5} =$

7. $7\frac{4}{9} + 2\frac{1}{6} =$

8. $2\frac{4}{5} + 3\frac{1}{6} =$

9. $5\frac{7}{10} + 2\frac{3}{4} =$

Problem Solving

Solve.

10. Jake biked $4\frac{5}{8}$ miles from the cottage to the beach and another $2\frac{1}{2}$ miles to work. How many total miles did he ride?

11. Aki played a computer game for $1\frac{5}{6}$ hours yesterday and played for $1\frac{3}{4}$ hours today. How long did she play in all?

12. Tina ran for $15\frac{3}{5}$ minutes one day and $17\frac{9}{10}$ minutes another day. How long did she run in all?

13. Sol left his house and spent $2\frac{1}{2}$ hours at the movies and $1\frac{3}{4}$ hours at the gym. How long was he gone?

14. The first act of the play took $1\frac{2}{3}$ hours. The second act lasted $2\frac{1}{4}$ hours. How long was the play?

15. You volunteered $3\frac{2}{5}$ hours last week and $12\frac{4}{6}$ hours this week. How many hours have you volunteered?

LESSON 12 Subtracting Fractions with Like Denominators

As with adding fractions, when you subtract fractions, you need to make sure they have a common denominator. If the fractions already have the same denominator, just subtract the numerators.

$$\frac{5}{6} - \frac{3}{6} = \frac{2}{6}$$

The basic idea is this:

$$\frac{\text{numerator}}{\text{denominator}} - \frac{\text{numerator}}{\text{denominator}} = \frac{\text{numerator} - \text{numerator}}{\text{denominator}}$$

When you subtract fractions, you subtract the numerators, but you <u>do</u> <u>not</u> subtract the denominators.

After you subtract, you must check to make sure that the fraction is in simplest form.

Example

Subtract. $\frac{9}{10} - \frac{3}{10}$

STEP 1 If both fractions have the same denominator, subtract the numerators.

$$\frac{9 - 3}{10} = \frac{6}{10}$$

STEP 2 If the result is an improper fraction, change it to a mixed number. In this case, $\frac{6}{10}$ is a proper fraction, so there is no need to rename it as a mixed number.

STEP 3 Simplify the fraction. $\frac{6}{10}$ is not in simplest form. Divide the numerator and denominator by 2.

$$\frac{6}{10} \div \frac{2}{2} = \frac{3}{5}$$

$$\frac{9}{10} - \frac{3}{10} = \frac{6}{10} = \frac{3}{5}$$

ON YOUR OWN

After lunch, Jana had $\frac{7}{8}$ of a pizza left. She gave $\frac{3}{8}$ of the pizza to her sister. What part of the pizza is left? Simplify if possible.

Practice

Building Skills

Subtract.

1. $\dfrac{5}{6} - \dfrac{1}{6} =$

2. $\dfrac{7}{8} - \dfrac{5}{8} =$

3. $\dfrac{4}{4} - \dfrac{1}{4} =$

4. $\dfrac{3}{5} - \dfrac{1}{5} =$

5. $\dfrac{8}{9} - \dfrac{5}{9} =$

6. $\dfrac{10}{10} - \dfrac{1}{10} =$

7. $\dfrac{11}{12} - \dfrac{7}{12} =$

8. $\dfrac{13}{16} - \dfrac{5}{16} =$

9. $\dfrac{20}{20} - \dfrac{13}{20} =$

Problem Solving

Solve.

10. Lorenzo lives $\frac{8}{10}$ of a mile from school. Angelo lives $\frac{4}{10}$ of a mile from school. How much farther does Lorenzo live from the school than Angelo?

11. Jesse read $\frac{1}{12}$ of a book. What fraction of the book does he have left to read?

12. Shandra had $\frac{5}{8}$ of a cup of oil. She used $\frac{3}{8}$ of a cup of oil in a recipe. How much oil does Shandra have left?

13. Malik has to weed $\frac{11}{12}$ of the garden in the backyard. He weeded $\frac{5}{12}$ of the garden on Saturday. What fraction of the garden does he have left to weed on Sunday?

14. Rachel bought an eight-slice pizza. She ate $\frac{1}{8}$ of the pizza. How much was left?

15. The sporting goods store sells exercise mats that are $\frac{7}{8}$ of an inch thick. Another store sells mats that are $\frac{5}{8}$ of an inch thick. How much thicker are the mats sold by the first store?

LESSON 13 Subtracting Fractions with Unlike Denominators

When you subtract fractions, their denominators must be the same. But what happens when you need to subtract fractions that have different denominators? You have to make sure that the fractions have the same denominator. You will need to find equivalent fractions. Equivalent fractions are fractions that name the same value. Then you simply subtract the numerators.

Example

Subtract. $\frac{2}{3} - \frac{1}{2}$

STEP 1 Find the least common denominator (LCD).
Find the LCD by listing the multiples of each denominator.
3: 3, 6, 9, 12
2: 2, 4, 6, 8, 10
The LCD of 2 and 3 is 6.

STEP 2 Write equivalent fractions with the LCD.
Remember, to get the equivalent fraction, you must multiply the numerator and denominator by the same number.

$$\frac{2}{3} \times \frac{2}{2} = \frac{4}{6}$$
$$\frac{1}{2} \times \frac{3}{3} = \frac{3}{6}$$

STEP 3 Subtract the numerators and keep the denominator the same.

$$\frac{4-3}{6} = \frac{1}{6}$$

STEP 4 Simplify the fraction if necessary.
In this case this step is not necessary because $\frac{1}{6}$ is in simplest form.
$$\frac{2}{3} - \frac{1}{2} = \frac{1}{6}$$

ON YOUR OWN

Shannon played a video game for $\frac{3}{4}$ of an hour. Lori played the same game for $\frac{2}{3}$ of an hour. How much longer did Shannon play? Simplify if possible.

Practice

Be sure your answer is in its simplest form.

Building Skills

Subtract.

1. $\dfrac{7}{10} - \dfrac{1}{2} =$

2. $\dfrac{2}{3} - \dfrac{1}{6} =$

3. $\dfrac{4}{5} - \dfrac{1}{10} =$

4. $\dfrac{1}{3} - \dfrac{1}{4} =$

5. $\dfrac{5}{9} - \dfrac{1}{3} =$

6. $\dfrac{7}{8} - \dfrac{3}{4} =$

7. $\dfrac{11}{12} - \dfrac{3}{8} =$

8. $\dfrac{15}{16} - \dfrac{3}{4} =$

9. $\dfrac{5}{9} - \dfrac{1}{6} =$

Problem Solving

Solve.

10. You walk $\frac{9}{10}$ of a mile from your home to the center of town. On the way back, you stop at a store that is $\frac{1}{3}$ of a mile from the town center. How far is the store from your home?

11. Larissa ran $\frac{1}{5}$ of a mile on Thursday and $\frac{7}{10}$ of a mile on Friday. How much farther did Larissa run on Friday than on Thursday?

12. Deion needed $\frac{7}{8}$ of a yard of material to make a banner. If he has $\frac{5}{6}$ of a yard, how much more does he need?

13. Anna ran $\frac{9}{10}$ of a mile. Lucia ran $\frac{3}{4}$ of a mile. How much farther did Anna run?

LESSON 14   Subtracting Mixed Numbers

One day at your summer job, you need to take time off for a doctor's appointment. Ordinarily, your workday is $8\frac{1}{2}$ hours long, but you are away for $1\frac{1}{4}$ hours. How long do you end up working that day?

In one way, subtracting mixed numbers is like adding mixed numbers: when you subtract, you subtract the whole numbers and fractions separately.

But sometimes you cannot subtract fractions as easily as you can add them. That is when you have to rename one of the mixed numbers.

Example

Subtract. $8\frac{1}{2} - 1\frac{1}{4}$

STEP 1 **Find the least common denominator (LCD) of the fractions.**
List the multiples of 2 and 4.
2: 2, 4, 6, 8, 10
4: 4, 8, 12, 16
The LCD of 2 and 4 is 4.

$$\frac{1}{2} \times \frac{2}{2} = \frac{2}{4}$$
$$\frac{1}{4} \times \frac{1}{1} = \frac{1}{4}$$

STEP 2 **Write equivalent fractions using the LCD.**

STEP 3 **If necessary, rename one of the wholes as a fraction. Add it to the fraction that is there already.**
Since you can subtract $\frac{1}{4}$ from $\frac{2}{4}$, you do not need to rename any other part of the mixed number.

STEP 4 **Subtract the whole numbers and the fractions.**
$8 - 1 = 7$

$$\frac{2}{4} - \frac{1}{4} = \frac{1}{4}$$

STEP 5 **Combine the new whole number and the new fraction to make a new mixed number or proper fraction. Simplify if necessary.**
The difference is $7\frac{1}{4}$. In this example, the difference is already in simplest form.

$$8\frac{1}{2} - 1\frac{1}{4} = 7\frac{1}{4}$$

$$7 + \frac{1}{4} = 7\frac{1}{4}$$

(ON YOUR OWN)

Jamie caught a fish that weighs $2\frac{3}{4}$ pounds. Her uncle caught a fish that weighs $3\frac{1}{2}$ pounds. How many more pounds does Jamie's uncle's fish weigh?

Practice

Building Skills

Subtract.

1. $3\frac{1}{3} - 1\frac{1}{2} =$

2. $4\frac{5}{8} - 2\frac{1}{4} =$

3. $5\frac{2}{3} - 1\frac{1}{2} =$

4. $4\frac{3}{4} - 2\frac{2}{3} =$

5. $3\frac{1}{4} - 1\frac{1}{2} =$

6. $6\frac{3}{8} - 2\frac{7}{8} =$

7. $8\frac{2}{5} - 1\frac{1}{3} =$

8. $3\frac{1}{6} - 1\frac{11}{12} =$

9. $12\frac{3}{8} - 6\frac{5}{6} =$

Problem Solving

Solve.

10. Mary Ellen swam for $12\frac{2}{3}$ minutes yesterday. The day before she swam for only $9\frac{1}{2}$ minutes. How much longer did she swim the second day?

11. Ms. Gutierrez had $2\frac{1}{3}$ pounds of chocolate for a party. Her guests ate $1\frac{1}{4}$ pounds. How many pounds of chocolate were left?

12. Renee used $2\frac{3}{4}$ rolls of film at the party and $4\frac{1}{2}$ rolls of film at the graduation ceremony. How many more rolls of film did she use at the graduation?

13. Mariana filled $2\frac{1}{3}$ bins for recycling. Eric filled $1\frac{1}{4}$ bins. How much more recycling material did Mariana have?

14. Claudio swam $3\frac{5}{8}$ lengths of the pool. Derrick swam $1\frac{1}{4}$ lengths of the pool fewer than Claudio did. How many lengths of the pool did Derrick swim?

15. Melissa painted $4\frac{5}{6}$ square feet of the wall. Althea painted $1\frac{3}{4}$ square feet less than Melissa did. How many square feet of the wall did Althea paint?

Name _____ Date _____

LESSON 15   Multiplying Fractions

Suppose you have $\frac{1}{2}$ dollar and your friend asks you for $\frac{1}{2}$ of your money. You would then have $\frac{1}{2}$ of a $\frac{1}{2}$ dollar, which is the same as 25 cents, or a quarter ($\frac{1}{4}$ of a dollar). You can write: $\frac{1}{2} \times \frac{1}{2} = \frac{1}{4}$.

When you multiply fractions, you multiply the numerators by each other and the denominators by each other. Remember to simplify your answer if necessary.

$$\frac{7}{8} \times \frac{3}{4} = \frac{7 \times 3}{8 \times 4} = \frac{21}{32}$$

Example

Multiply. $\frac{7}{8} \times \frac{4}{5}$

STEP 1 Multiply numerator by numerator and denominator by denominator.

$$\frac{7 \times 4}{8 \times 5} = \frac{28}{40}$$

STEP 2 Make sure the answer (the product) is in simplest form. Remember, when you simplify a fraction, you look for a number that will divide evenly into both parts of the fraction. You can divide each part of the fraction by 4.

$$\frac{28}{40} \div \frac{4}{4} = \frac{7}{10}$$

$$\frac{7}{8} \times \frac{4}{5} = \frac{28}{40} = \frac{7}{10}$$

(ON YOUR OWN)

Helene painted $\frac{2}{3}$ of a wall white and $\frac{1}{2}$ of the wall red. What part of the wall was painted both red and white?

Practice

Building Skills

Multiply.

1. $\dfrac{5}{8} \times \dfrac{16}{25} =$

2. $\dfrac{3}{4} \times \dfrac{1}{3} =$

3. $\dfrac{2}{5} \times \dfrac{1}{6} =$

4. $\dfrac{2}{9} \times \dfrac{3}{5} =$

5. $\dfrac{5}{6} \times \dfrac{4}{5} =$

6. $\dfrac{8}{9} \times \dfrac{3}{4} =$

7. $\dfrac{15}{16} \times \dfrac{1}{10} =$

8. $\dfrac{6}{7} \times \dfrac{14}{15} =$

9. $\dfrac{9}{16} \times \dfrac{8}{15} =$

Problem Solving

Solve.

10. Ian's garden fills $\frac{2}{3}$ of his backyard. He plans to plant flowers in $\frac{1}{2}$ of the garden. How much of the backyard will be for flowers?

11. Andres has $\frac{1}{4}$ of a watermelon. He wants to give $\frac{1}{2}$ of this watermelon to his sister. What fraction of the watermelon is Andres going to give to his sister?

12. Rebecca's recipe calls for $\frac{4}{5}$ of a cup of flour. She plans to make $\frac{1}{4}$ of the recipe. How much flour will she use?

13. Joey ran $\frac{9}{10}$ of a mile. Juan ran $\frac{2}{3}$ as far as Joey. How far did Juan run?

14. The top of Elaine's desk measures 1 yard by $\frac{2}{3}$ of a yard. What is the area of the top of her desk? [Hint: area = length × width]

15. A floor-cleaning solution is made using $\frac{1}{2}$ cup of ammonia for every 3 gallons of water. How much ammonia would you need if you were making only one gallon of floor cleaner?

LESSON 16 Multiplying Mixed Numbers

Multiplying mixed numbers is similar to what you just did to multiply fractions. To multiply mixed numbers you:

- change the mixed numbers to improper fractions;
- multiply the fractions;
- rename the answer as a mixed number.

You can estimate your answer by rounding to the whole numbers and multiplying them. This gives you a number that is close to the actual product.

For example: $5\frac{1}{3} \times 2\frac{1}{2}$

Round the whole numbers: $5 \times 3 = 15$

The product should be close to 15.

Example

Multiply. $7\frac{1}{2} \times 3\frac{1}{4}$

STEP 1 Rename the mixed numbers as improper fractions.

$$7\frac{1}{2} = \frac{15}{2} \quad 3\frac{1}{4} = \frac{13}{4}$$

STEP 2 Multiply numerator by numerator, denominator by denominator.

$$\frac{15 \times 13}{2 \times 4} = \frac{195}{8}$$

STEP 3 Simplify the fraction and/or rename it as a mixed number.

$$\frac{195}{8} = 24\frac{3}{8}$$

STEP 4 Estimate the product to see if your result is reasonable.
$8 \times 3 = 24$

Your estimate is close to your answer.

$$7\frac{1}{2} \times 3\frac{1}{4} = 24\frac{3}{8}$$

$$7\frac{1}{2} \times 3\frac{1}{4} = 24\frac{3}{8}$$

ON YOUR OWN

Jorge's room measures $12\frac{1}{2}$ feet by $9\frac{2}{5}$ feet. What is the area of his room? (Hint: area = length × width)

Practice

Building Skills

Multiply.

1. $3\frac{1}{3} \times 4\frac{1}{2} =$

2. $1\frac{1}{4} \times 2\frac{2}{5} =$

3. $3\frac{1}{8} \times 4\frac{4}{5} =$

4. $5\frac{1}{4} \times 2\frac{4}{7} =$

5. $6\frac{1}{2} \times 3\frac{1}{5} =$

6. $3\frac{3}{4} \times 1\frac{1}{15} =$

7. $4\frac{3}{8} \times 4\frac{4}{7} =$

8. $3\frac{2}{3} \times 1\frac{1}{11} =$

9. $2\frac{13}{16} \times 4\frac{4}{9} =$

Problem Solving

Solve.

10. Jake's cubicle at his magazine internship measures $3\frac{1}{2}$ feet by $5\frac{1}{3}$ feet. What is the area of the cubicle?
(*Hint:* area = length × width)

11. Rosa's formula uses $2\frac{1}{2}$ cups of water. If she wants to make $2\frac{1}{2}$ times the formula, how much water does she need?

12. Dana swims $5\frac{1}{2}$ laps in 2 minutes. Justine can swim $1\frac{1}{2}$ times as far in the same amount of time. How far will Justine swim in 2 minutes?

13. Terrell needs 6 pieces of cloth that are each $4\frac{1}{2}$ feet long for decorations for the dance. How much material should he buy?

14. The size of a note card is $2\frac{3}{4}$ inches by $4\frac{1}{3}$ inches. What is the area of the card in square inches? (*Hint:* area = length × width)

15. Leila is running a road race. Her average speed is $7\frac{1}{3}$ miles per hour. How long is the race if Leila finishes in $1\frac{1}{2}$ hours?

LESSON 17   Dividing Fractions

Division is the opposite of multiplication. You can apply the skills you have learned for multiplying fractions to dividing fractions. When you divide fractions, you have to remember to **invert** the fraction to the right of the division sign before you multiply.

Example

Divide. $\dfrac{9}{10} \div \dfrac{3}{4}$

STEP 1 Invert the fraction to the right of the division sign.

$$\boxed{\dfrac{3}{4} \rightarrow \dfrac{4}{3}}$$

STEP 2 Change the division sign ($\div$) to a multiplication sign ($\times$).

$$\boxed{\dfrac{9}{10} \times \dfrac{4}{3}}$$

STEP 3 Multiply.

$$\boxed{\dfrac{9}{10} \times \dfrac{4}{3} = \dfrac{9 \times 4}{10 \times 3} = \dfrac{36}{30}}$$

STEP 4 Simplify the fraction and change to a mixed number if necessary.

$$\dfrac{9}{10} \div \dfrac{3}{4} = 1\dfrac{1}{5}$$

$$\boxed{\dfrac{36}{30} = 1\dfrac{6}{30} = 1\dfrac{1}{5}}$$

ON YOUR OWN

Jennie cut a cake into 6 slices and ate some. She has $\dfrac{2}{3}$ of the cake left. How many slices does she have left? (Hint: $\dfrac{2}{3} \div \dfrac{1}{6}$)

Practice

Invert the fraction to the right of the division sign.

Building Skills

Divide.

1. $\dfrac{5}{6} \div \dfrac{1}{3} =$

2. $\dfrac{3}{4} \div \dfrac{1}{2} =$

3. $\dfrac{9}{10} \div \dfrac{1}{4} =$

4. $\dfrac{7}{8} \div \dfrac{1}{3} =$

5. $\dfrac{4}{9} \div \dfrac{2}{3} =$

6. $\dfrac{5}{6} \div \dfrac{7}{8} =$

7. $\dfrac{15}{16} \div \dfrac{3}{16} =$

8. $\dfrac{21}{25} \div \dfrac{7}{10} =$

9. $\dfrac{4}{5} \div \dfrac{1}{6} =$

Problem Solving

Solve.

10. Lakesha has $\frac{3}{4}$ of a cup of walnuts for cookies. If she plans to add $\frac{1}{3}$ of a cup per dozen cookies, how many dozen cookies can she make?

11. Anna is cutting pieces of wood for a birdhouse. If she has a board $\frac{3}{4}$ of a yard long and wants to cut pieces that are $\frac{1}{8}$ of a yard long, how many pieces will she get?

12. Franco has $\frac{5}{8}$ of a pie left. How many $\frac{1}{16}$ pieces can he cut from what he has now?

13. How many pieces of rope $\frac{1}{12}$ of a yard long can you cut from a piece of rope that measures $\frac{2}{3}$ of a yard?

LESSON 18 Dividing Mixed Numbers

When you divide mixed numbers, you must first change the mixed numbers to improper fractions. Then you divide the fractions the same way as you did in the last lesson. You must invert the fraction to the right of the division sign.

Example

Divide. $3\frac{1}{5} \div 1\frac{13}{15}$

STEP 1 Change the mixed numbers to improper fractions.

$$\frac{16}{5} \div \frac{28}{15}$$

STEP 2 Invert the fraction to the right of the division sign.

$$\frac{28}{15} \rightarrow \frac{15}{28}$$

STEP 3 Change the division sign ($\div$) to a multiplication sign ($\times$).

$$\frac{16}{5} \times \frac{15}{28} =$$

STEP 4 Multiply.

$$\frac{16}{5} \times \frac{15}{28} = \frac{15 \times 16}{5 \times 28} = \frac{240}{140}$$

STEP 5 Simplify and change the answer to a mixed number if necessary.

$$3\frac{1}{5} \div 1\frac{13}{15} = 1\frac{5}{7}$$

$$\frac{240}{140} = 1\frac{10}{14} = 1\frac{5}{7}$$

ON YOUR OWN

Melissa rode her bike $23\frac{1}{3}$ miles in $2\frac{1}{2}$ hours. At that rate, how far did she ride in 1 hour?

Practice

Invert the fraction to the right of the division sign.

Building Skills

Divide.

1. $3\frac{3}{4} \div 2\frac{1}{2} =$

2. $5\frac{1}{3} \div 1\frac{1}{6} =$

3. $1\frac{9}{10} \div 3\frac{1}{5} =$

4. $4\frac{1}{2} \div 3\frac{1}{3} =$

5. $3\frac{3}{4} \div 1\frac{2}{5} =$

6. $4\frac{1}{6} \div 1\frac{7}{8} =$

7. $1\frac{5}{12} \div 1\frac{5}{6} =$

8. $2\frac{2}{25} \div 1\frac{1}{10} =$

9. $8\frac{1}{3} \div 2\frac{1}{5} =$

Problem Solving

Solve.

10. Akiko has a recipe for a large omelet that requires $2\frac{3}{4}$ cups of milk. She wants to make an omelet that is $1\frac{1}{2}$ times smaller than the recipe. How many cups of milk does Akiko need for her omelet?

11. Jorge has a pine board that is $12\frac{3}{8}$ feet long. He wants to cut it into $2\frac{1}{3}$-foot pieces. How many pieces, including the fraction of the leftover piece, will Jorge cut?

12. Martha is cutting rope into pieces for a craft project. The rope was $6\frac{1}{4}$ feet long, and there are $2\frac{1}{2}$ pieces. How long is each piece?

13. Mandy is using a $1\frac{1}{3}$-pint scoop to scoop water out of a large pot. The pot has 16 pints in it. How many scoops will it take to empty the pot?

LESSON 19 Ratios

You probably have heard people say things like *The ratio of boys to girls in the class is 12 to 11* or *On a piano, the ratio of black keys to white keys is 5 to 7*.

A **ratio** compares two amounts or numbers. Like fractions, ratios must be written in *simplest form*.

Ratios can be written

as a fraction	with a colon	in words
$\dfrac{1}{2}$	1 : 2	1 to 2

Example

This basketball team has 9 players.

Look at the uniforms. Write a ratio comparing the number of uniforms that have odd numbers to the number of uniforms that have even numbers.

STEP 1 Identify the numbers that you need to compare.

odd numbers	even numbers
6 uniforms	3 uniforms

STEP 2 Write a ratio.
odd to even 6 to 3 or $\dfrac{6}{3}$

STEP 3 Write the ratio in simplest form.
Reduce $\dfrac{6}{3}$ to simplest form by dividing.
Use the GCF (greatest common factor).

$$\frac{6}{3} \div \frac{3}{3} = \frac{2}{1}$$

In simplest form, the ratio is
2 to 1, 2 : 1, or $\frac{2}{1}$.

> Finding the GCF of 6 and 3:
> - Factors of 6 are 1, 2, ③, 6
> - Factors of 3 are 1, ③
>
> 3 is the GCF of 6 and 3

ON YOUR OWN

There are 24 students in the school musical. Ten of the students are seniors. What is the ratio of students who are seniors to students who are *not* seniors? Simplify if possible.

Practice

Building Skills

Write a ratio in three ways. Reduce to simplest form.

There are 18 students on the school bus. There are 5 seniors, 2 juniors, 6 sophomores, and 5 freshmen.

1. What is the ratio of juniors to freshmen?

2. What is the ratio of juniors to seniors?

3. What is the ratio of sophomores to seniors?

4. What is the ratio of sophomores to all bus riders?

Use the grouping of letters to write each ratio. Reduce the ratio to simplest form.

C A C A C B C
A A B C B C A

5. What is the ratio of As to Bs?

6. What is the ratio of As to Cs?

7. What is the ratio of Bs to Cs?

8. What is the ratio of Cs to As *and* Bs?

Problem Solving

Write a ratio.

9. Write a ratio to compare the number of letters in the word *Picasso* with the number of letters in the word *Shakespeare*.

10. Sarah swims 35 laps. Kevin swims 25 laps. Write a ratio comparing the number of laps that Kevin swims to the number of laps that Sarah swims.

11. Ethan and Ben played against each other in a trivia game. Ethan scored 8 points and Ben scored 12 points. Write the ratio, in simplest form, that compares Ethan's score to Ben's score.

12. In a school spirit contest, Pat's team of 11 people had 7 of the winning costumes. Write a ratio that compares the number of winning costumes to the total number of people on the team.

LESSON 20 Rates

You have learned about ratios. Now you are ready to learn about a special kind of ratio called a rate. A **rate** is a ratio that compares two amounts measured in *different* kinds of units. Here are examples:

250 *miles* in 4 *hours*   Inez drove at a rate of 250 miles in 4 hours.

200 *sit-ups* in 30 *minutes*   Alex can do 200 sit-ups in 30 minutes.

200 *dollars* for 1 *day*   Lee charges clients $200 for one day of work.

You use and see rates every day. When you go shopping, you might see a sign similar to the one above.

Example

Use the sale sign above. Write the sale price for CDs as a rate. Express the rate in simplest form.

STEP 1 Write the numbers you are comparing as a fraction.
The rate is $\dfrac{24 \text{ dollars}}{3 \text{ CDs}}$.

STEP 2 Express the rate in simplest form.
Divide the numerator and denominator by their GCF, 3.

$= \dfrac{24 \text{ dollars} \div 3}{3 \text{ CDs} \div 3}$

$= \dfrac{8 \text{ dollars}}{1 \text{ CD}}$

CDs are selling at the rate of $8 per CD.

ON YOUR OWN

Marcus can flip a coin 140 times in 4 minutes. At what rate does he flip a coin?

Practice

To simplify, find the GCF.

Building Skills

Write each rate in simplest form.

1. 10 miles in 6 hours

2. 20 dollars for 6 books

3. 9 free throws in 24 attempts

4. 50 liters in 4 minutes

5. $32 earned in 4 hours

6. 5 tickets for $80

7. 6 cups of flour for 4 eggs

8. $300 for 4 tires

9. 10 laps in 5 minutes

Problem Solving

Write each rate in simplest form.

10. Abdul drove 80 miles in 2 hours.

11. In 4 hours, Carmen used 10 rolls of film.

12. Jerome swims 15 laps in 20 minutes.

13. The Tanaka family spent $105 for a 14-day travel pass.

14. Maria's dog chased 30 squirrels in 4 hours.

15. Olivia used 30 gallons of gas on a 510-mile road trip.

Name _____ Date _____

LESSON 21 Learning About Unit Rates

You have learned that a rate is a ratio comparing different kinds of units. You also know that rates can be simplified like fractions. Now you will learn about a very useful kind of rate called a *unit rate*.

A **unit rate** is the rate for *one* unit of a quantity. Here are some examples:

$$\frac{45 \text{ miles}}{\text{hour}} \qquad \frac{65 \text{ words}}{\text{minute}} \qquad \frac{25 \text{ students}}{\text{teacher}}$$

The denominator for any unit rate is 1—as in *1* hour, *1* minute, or *1* teacher.

A unit rate that tells the price per unit is called a **unit price.**

$$\frac{\$3.00}{\text{box}} \qquad \frac{\$0.25}{\text{can}} \qquad \frac{\$2.79}{\text{fluid ounce}}$$

When you read nutrition facts on food packages, you are reading unit rates.

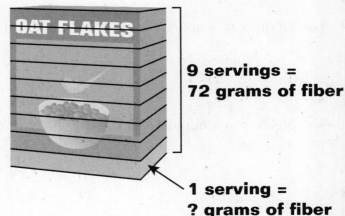

9 servings = 72 grams of fiber

1 serving = ? grams of fiber

Example

A box of cereal contains 9 servings and a total of 72 grams of fiber. How many grams of fiber are in one serving?
(One serving is 1 unit.)

STEP 1 Write the numbers as a fraction.

$$\frac{72 \text{ grams}}{9 \text{ servings}}$$

STEP 2 Divide.

$$\frac{72 \text{ grams}}{9 \text{ servings}} = \frac{8 \text{ grams}}{1 \text{ serving}}$$

There are 8 grams of fiber in one serving.

ON YOUR OWN

Jamal drove 125 miles in $2\frac{1}{2}$ hours. How many *miles per hour* did he drive?

Practice

Building Skills

Write the unit rate or unit price.

1. 60 miles in 4 hours

2. $45 for 5 hours

3. 200 words in 4 minutes

4. 88 miles in 2 hours

5. 6 cans of juice for $1.80

6. 36 grams of fat in 6 servings

7. 3,000 meters in 10 minutes

8. 144 miles on 6 gallons

9. 90 pages in 180 minutes

Problem Solving

Write the unit rate in simplest form.

10. Twelve small bottles of juice sell for $4.80.

11. Barak reads 60 pages in 30 minutes.

12. Rob can type 500 words in 10 minutes.

13. A pizza that costs $9 has 6 slices.

14. The Ramos family pays $350 for 4 rooms at an inn.

15. Five-pound watermelons are on sale for $4 each. Find the price per pound.

LESSON 22 Proportions

You know that ratios make comparisons. Two ratios that make the same comparison are called **equivalent ratios**.

A statement that shows two ratios as equal is called a **proportion**.

To check to see if the ratios are equal, **cross multiply**. If the **cross products** are equal, they form a proportion.

(1) × (2) $3 \times 2 = 6$
(3) × (6) $1 \times 6 = 6$

Because $6 = 6$ the ratios are a proportion.

equivalent—when two things are equal

$$\frac{2 \text{ basketballs}}{6 \text{ basketballs}} = \frac{1 \text{ basketball}}{3 \text{ basketballs}}$$

Equivalent ratios are proportions.

Example

On a test, Suki got 6 answers right out of 9. Jim got 8 answers right out of 10. Are the students' scores proportional?

STEP 1 Set up the two ratios. Be sure that the numbers in the ratios are set up in correct order.

Suki	Jim
$\frac{6}{9}$	$\frac{8}{10}$

STEP 2 Find the cross products.

$$\frac{6}{9} \times \frac{8}{10}$$

$6 \times 10 = 60$
$9 \times 8 = 72$

STEP 3 Compare the cross products.

$60 \neq 72$, so the ratios are *not* a proportion.
Suki and Jim's scores are <u>not</u> proportional.

ON YOUR OWN

This summer 8 out of 12 people took a vacation. Last summer 3 out of 4 people took a vacation. Are these ratios proportional?

Practice

Building Skills

Write = or ≠ between the ratios.

1. 2 : 3   12 : 18

2. $\dfrac{3}{8}$ $\dfrac{6}{18}$

3. $\dfrac{2}{5}$ $\dfrac{8}{20}$

4. 6 : 8   12 : 16

5. 4 : 9   12 : 26

6. $\dfrac{3}{7}$ $\dfrac{14}{6}$

7. $\dfrac{7}{2}$ $\dfrac{21}{6}$

8. 9 : 5   27 : 9

9. $\dfrac{2}{8}$ $\dfrac{20}{80}$

Problem Solving

Use proportions to solve these problems.

10. Jim gets paid $40 for 4 hours of work. Carmela gets paid $60 for 6 hours of work. Do Jim and Carmela get paid at the same rate?

11. Anna saves $3 for every $10 she earns. Patrick saves $5 for every $20 he earns. Do the two save money at the same rate?

12. Henry completed 6 of the 15 passes he attempted. Giovanni completed 9 of his 21 passes. Did the two quarterbacks complete passes at the same rate?

13. Jackie can type 50 words in a minute. Her brother can type 200 words in 4 minutes. Are their typing speeds the same?

14. Keonta can walk half a mile in ten minutes. Abigail walks one mile in 25 minutes. Do Keonta and Abigail walk at the same rate?

15. Emily answered 8 of 10 questions correctly on a test. On a different test, Yuki answered 15 of 20 questions correctly. Did Emily and Yuki have the same ratio of correct answers?

LESSON 23 Using Cross Products

You know that you can use cross products to find whether two ratios form a proportion. You can also use cross products to find a missing number in a proportion.

It doesn't matter how you set up the ratios as long as the same categories are on top and the same categories are on the bottom.

Example

Read the clipping. If Keisha keeps getting hits at this same rate, how many hits will she get in 60 at-bats?

STEP 1 Set up a proportion.
You do not know how many hits Keisha will get in 60 at-bats. Use a letter, such as n, to substitute for the unknown number.

$$\frac{\text{hits}}{\text{at-bats}} \rightarrow \frac{4}{12} = \frac{n}{60}$$

STEP 2 Cross multiply.

$$\frac{4}{12} = \frac{n}{60}$$

$12 \times n = 4 \times 60$
$12n = 240$

STEP 3 Solve for n.
Divide both sides by 12 because n is next to 12.
$12n = 240$
$$\frac{12n}{12} = \frac{240}{12}$$
$n = 20$
$$\frac{4}{12} = \frac{20}{60}$$

Keisha would get 20 hits in 60 at-bats.

> ## Teen Slugger Leads the Way
>
> Keisha Jones is off to a big start this season with 4 hits in her first 12 at-bats. Her 9th inning round-tripper gave the Tigers their third straight victory.

ON YOUR OWN

One photo is 6 inches long and 4 inches wide. You can enlarge the photo so that it will be 18 inches long. How wide will the photo be?

Practice

Building Skills

Solve each proportion.

1. $\dfrac{3}{4} = \dfrac{n}{12}$

2. $\dfrac{4}{n} = \dfrac{6}{9}$

3. $\dfrac{8}{16} = \dfrac{n}{4}$

4. $\dfrac{x}{30} = \dfrac{5}{6}$

5. $\dfrac{18}{30} = \dfrac{n}{10}$

6. $\dfrac{40}{n} = \dfrac{8}{5}$

7. $\dfrac{5}{25} = \dfrac{4}{x}$

8. $\dfrac{3}{8} = \dfrac{21}{y}$

9. $\dfrac{y}{20} = \dfrac{11}{4}$

Problem Solving

Use proportions to solve these problems.

10. LaToya gets paid by the hour. During a 20-hour week, LaToya earns $120. If she works only 15 hours, how much does she earn?

11. Joe is a photographer. For every 15 rolls of film he shoots, 10 are black and white. If Joe uses 57 rolls of film, how many rolls are in black and white?

12. Luisa swims 35 laps in 25 minutes. How many laps will she swim in 10 minutes?

13. Darnell walks 140 minutes in 4 days. At this rate, how many minutes will he walk in 6 days?

14. Dan scored 80 points in 5 basketball games. At this rate, how many points do you expect him to score in 20 games?

15. Jay buys 60 square feet of tile for $324. Later, he buys another 20 square feet of tile. How much does the extra tile cost?

LESSON 24 Writing Proportions

You have solved proportions by finding cross products. The numbers used in proportions are called **terms**. If you know three of the four terms of the proportion, you can solve for the missing term.

Look at the nutrition facts below.

Nutrition Facts	
Serving Size 2 crackers (13g)	
Servings Per Container About 20	
Amount Per Serving	
Calories 70	Calories from Fat 30
	%Daily Value*
Total Fat 3g	5%
Saturated Fat 1.5g	8%

You just ate 7 crackers. How can you find the total number of calories in 7 crackers? The number of calories in each cracker stays the same. So the ratio of calories to crackers must be the same, too. Therefore, you can use the idea of equivalent ratios to solve the problem.

Example

Look at the nutrition facts above. How many calories are in 7 crackers?

STEP 1 Identify the information you have.
- A serving size is 2 crackers.
- There are 70 calories in 1 serving.
- You ate 7 crackers.

STEP 2 Write a proportion to show the comparisons.
One serving is 2 crackers. There are 70 calories in 2 crackers. Use the letter n to stand for the unknown number of calories in 7 crackers.

$$\frac{70 \text{ calories}}{2 \text{ crackers}} = \frac{n \text{ calories}}{7 \text{ crackers}} \leftarrow \text{Equivalent ratios that form a proportion must compare the } \textit{same} \text{ units in the } \textit{same} \text{ order.}$$

STEP 3 Solve for n.

$$\frac{70}{2} = \frac{n}{7}$$
$$70 \times 7 = 2 \times n$$
$$490 = 2n$$
$$245 = n$$

There are 245 calories in 7 crackers.

ON YOUR OWN

Julio can type 90 words in 2 minutes. He needs to take a 5-minute typing test in order to get a job. If Julio types at his usual rate, how many words will he type on this test?

Practice

© Harcourt Achieve Inc. All rights reserved.

> Compare the same units in the same order.

Building Skills

Set up a proportion and solve each problem.

1. You work 3 hours every 2 days. How many hours do you work in 6 days?

2. You read 42 pages in 3 hours. How many hours will it take you to read 147 pages?

3. It takes you 20 minutes to outline 5 pages in your history book. How much time will it take you to outline a 30-page chapter?

4. There were 7 field goals made in 14 attempts. How many goals should be made in 8 attempts?

5. You attend club meetings 18 days per month. How many meetings do you attend in 9 months?

6. A bus driver picks up 40 people in 5 hours. How many people does she pick up in 40 hours?

7. You can type 171 words in 3 hours. How many words do you type per hour?

8. Two out of every 7 students who apply to work in the library get a job. If 30 students have a library job, how many students applied?

Problem Solving

Use a proportion to solve each problem.

9. Eric earns $390 during a 30-hour week. How much does he earn per hour?

10. There are 360 calories in 6 servings. How many calories are in a single serving?

11. Kim runs 8 laps in 20 minutes. At that rate, how many laps could he run in 60 minutes?

12. Lisa made 8 cell phone calls in 12 minutes. At that rate, how many calls will she make in 30 minutes?

13. You can finish 36 math problems in 1 hour. How many problems would you finish in 30 minutes?

14. A recipe for carrot salad makes 12 servings. The recipe calls for 2 cups of grated carrots. You have only half a cup. How many servings can you make?

LESSON 25 Problem Solving Using Proportions

You know that you can use proportions to solve everyday problems. When you write a proportion, think about the following:

- All rates should stay the same.

- Make sure that both ratios compare the same units in the same order.

Look at the grocery-store ad at the right. If 6 limes cost $0.90, then twice as many limes cost twice as much. This proportion shows that the price for each lime is the same.

$$\frac{6 \text{ limes}}{\$0.90} = \frac{12 \text{ limes}}{\$1.80}$$

TODAY ONLY

Fruit Sale!

3 pints of blueberries	$5.00
2 honeydew melons	$5.00
1 dozen oranges	$3.00
6 limes	$0.90

Example

Look at the ad above. You are making blueberry muffins. You have $15 to buy blueberries. How many pints of blueberries can you buy?

STEP 1 Identify the information you have.
- 3 pints of blueberries for $5
- $15 to spend

STEP 2 Write a proportion to show the comparisons.
Compare the *same* units in the *same* order. One way is to compare blueberry pints for $5 with blueberry pints for $15.

$$\frac{3 \text{ pints}}{\$5} = \frac{n \text{ pints}}{\$15}$$

STEP 3 Solve the proportion.

$$\frac{3}{\$5} = \frac{n}{\$15}$$

$$\frac{45}{5} = \frac{5n}{5} \leftarrow \text{Divide both sides by 5 to find } n.$$

$$n = 9$$

For $15, you can buy 9 pints of blueberries.

ON YOUR OWN

Look at the ad again. You need to buy some honeydew melons. How much will 5 melons cost?

Practice

Problem Solving

Use a proportion to solve each problem.

1. If 3 onions weigh 1 pound, how much do 12 onions weigh?

2. If 5 tropical fish cost $6, how much will 10 tropical fish cost?

3. You buy a package of pens for $4.80. There are 12 pens in the package. How much would 10 pens cost?

4. Roses are on sale 6 for $10. How many roses could you buy for $15?

5. You guess that you can bike 35 miles in 2 hours. At that rate, how far could you bike in 6 hours?

6. A recipe for applesauce serves 8 people. You need three-quarters of a cup of water. Suppose you are making the recipe for 12 people. How much water will you need?

7. You volunteer to make calls for your favorite charity. You can make 24 calls in 2 hours. At that rate, how many calls can you make in half an hour?

8. One dozen bananas weigh 3 pounds. How much do 2 bananas weigh?

9. The school band plays for 20 minutes at a pep rally. They play 10 songs. If all the songs are the same length, how many songs do they play in 5 minutes?

10. You reach into a large bag of marbles and pull out a handful of 20. Four of them are black. You put the marbles back. There are a total of 200 marbles in the bag. Based on your handful, how many of the marbles in the bag will be black?

LESSON 26 Understanding Percents

At a discount store, Mia purchased this ring on sale.

What percent of a dollar did she spend?

You know that a dollar is 100 cents. When you compare a number to 100, it is called a **percent**. Percent (%) means *per hundred*. You can write the ratio $\frac{39}{100}$ as 39%.

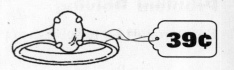

Mia spent 39 cents or $\frac{39}{100}$ or 39% of her dollar.

Some percents are easy to understand and remember.

> 100% of something means *all of it*.
> 0% of something is *none* of it.

dollar	dime	quarter
$1.00	$0.10	$0.25
100¢	10¢	25¢
$\frac{100}{100}$	$\frac{10}{100}$	$\frac{25}{100}$
100%	10%	25%

Example

You pay $100 total to have your car repaired. Of that total, $15 is for the needed car part. The cost of the part is what percent of the repair bill?

STEP 1 Write a ratio.

$$\frac{\$15}{\$100}$$

STEP 2 Write this ratio as a percent.

$$\frac{15}{100} = 15\%$$

$15 is 15% of the bill.

ON YOUR OWN

One hundred students tried out for the school play. Twenty-two of the students got a part in the play. Write this ratio as a percent.

Practice

Building Skills

Write each as a percent.

1. 44¢ as part of a dollar

2. 58 out of 100 rock musicians

3. 91 out of 100 soccer players

4. $30 out of $100

5. 75¢ as part of a dollar

6. 85 wins in 100 games

7. 66 out of 100 singers

8. 92 out of 100 questions answered correctly

9. 53 girls in a class of 100 students

10. 14 letters lost out of 100 sent

Problem Solving

Find the percent.

11. Carlos has a dollar. He spends 50¢ on a pen. What percent of a dollar does he spend?

12. A salesperson earns a $17 bonus each time she sells a $100 item. The bonus is what percent of the cost of the item?

13. In the United States, some people pay $0.23 out of every $1.00 they make for taxes. What percent of $1.00 do these people pay in taxes?

14. A local softball team won 46 of the 100 games it played. What percent of games did the team *win*?

LESSON 27 Converting Fractions to Percents

In the last lesson, you learned how to change a fraction to a percent when the denominator is 100. In fact, *you can write any fraction as a percent.*

To change a fraction to a percent you can:

- multiply by 100%; or

$$\frac{3}{4} \times 100\% = \frac{3}{4} \times \frac{\overset{25}{\cancel{100}}}{1} = \frac{75}{1} = 75\%$$

- divide, move the decimal 2 places, and write the % sign.

$$\frac{3}{4} = 4\overline{)3.00} = 75\%$$
$$\phantom{\frac{3}{4} = 4)}\underline{-28}$$
$$\phantom{\frac{3}{4} = 4)-}20$$
$$\phantom{\frac{3}{4} = 4)-}\underline{20}$$

Example

Ahmad got 6 answers correct out of 8 questions on a math quiz. What percent did Ahmad get correct?

STEP 1 Set up the fraction.

$$\frac{6}{8}$$

STEP 2 Multiply OR divide.

$$\frac{6}{8} \times 100\% = \frac{\overset{3}{\cancel{6}}}{\underset{2}{\cancel{8}}} \times \frac{\overset{25}{\cancel{100}}}{1} = \frac{75}{1} = 75\% \quad \text{OR} \quad \frac{6}{8} = 8\overline{)6.00} = 75\%$$
$$\underline{-56}$$
$$40$$
$$\underline{40}$$

Ahmed scored 75%.

ON YOUR OWN

Germaine played a video game. In the game, she hit 12 of the 20 targets. What percent of the targets did she hit?

Practice

Building Skills

Write each fraction as a percent.

1. $\dfrac{4}{5}$

2. $\dfrac{9}{10}$

3. $\dfrac{11}{20}$

4. $\dfrac{3}{50}$

5. $\dfrac{3}{8}$

6. $\dfrac{22}{100}$

7. $\dfrac{64}{200}$

8. $\dfrac{14}{20}$

9. $\dfrac{80}{400}$

Problem Solving

Solve.

10. There are 20 members in the school chorus. Seventeen had never been in the chorus before. What percent of the members had never been in the chorus before?

11. To pass the written part of the driver's test in Nevada, you need to answer 40 out of 50 questions correctly. What percent of questions must be answered correctly?

12. There are 40 students helping with after-school tutoring. Twenty-eight of the students are juniors and seniors. What percent are juniors and seniors?

13. A survey shows that $\dfrac{7}{8}$ of mall shoppers use discount coupons. What percent of mall shoppers use discount coupons?

14. Twenty-four hikers out of 400 hikers got poison ivy. What percent of the hikers got poison ivy?

15. A skateboarder cleanly performs a new move 16 out of 80 tries. What percent of the time does she cleanly perform the move?

LESSON 28 Converting Percents to Fractions

Writing a percent as a fraction is easier than writing a fraction as a percent.

You only need to remember how to simplify a fraction.

To write any percent as a fraction:

- write the percent as a fraction with a *denominator* of 100;
- simplify the fraction if possible.

> Remember, percent means *per hundred*. So any percent can be written in the fraction form $\frac{n}{100}$.

Example

Thirty percent of the students attend a driver's ed class after school. What fraction of the students attends driver's ed after school?

STEP 1 Write the percent as a fraction with a denominator of 100.

$$30\% = \frac{30}{100}$$

STEP 2 Simplify the fraction if possible. Divide the numerator and denominator by their greatest common factor (GCF).
Factors of 30 are 1, 2, 3, 5, 6, **10**, 15, 30.
Factors of 100 are 1, 2, 4, 5, **10**, 20, 25, 50, 100.
The GCF of 30 and 100 is 10.

$$\frac{30}{100} \rightarrow \frac{30 \div 10}{100 \div 10} \rightarrow \frac{3}{10}$$

$\frac{3}{10}$ of the students attend driver's ed after school.

ON YOUR OWN

In this year's graduating class, 85% of the students plan to go to college. What fraction of the students plans to go to college? Simplify if possible.

Practice

Building Skills

Simplify the fraction.

Write each percent as a fraction. Simplify if possible.

1. 55%

2. 95%

3. 15%

4. 44%

5. 83%

6. 28%

7. 8%

8. 34%

9. 68%

Problem Solving

Solve.

10. Forty-five percent of the teachers at your school can name at least one popular recording artist. What fraction of the teachers is this?

11. On a game show, one contestant correctly answered 43% of the questions. What fraction of the questions did she answer correctly?

12. Only 15% of the school clean-up committee members are new. What fraction of the members is new?

13. In a survey, 68% of students said that they collect trading cards. What fraction of the students collects trading cards?

14. At the 2002 Winter Olympics, 29% of the medals won by the United States were gold medals. What fraction of the medals won were gold medals?

15. Maureen is saving 28% of her pay for a class trip to the beach. What fraction of her pay is she saving for the trip?

LESSON 29 Converting Decimals to Percents

A survey of high-school students found that 0.74 favor having a student lounge. What percent of those surveyed favor having a student lounge?

You can write *any* decimal as a percent. Here are the steps to write 0.74 as a percent.

- Multiply the decimal by 100. This moves the decimal point 2 places to the *right*.

  $0.74 \times 100 = 074 = 74$

- Write the percent sign.

  $0.74 = 74\%$

74% of the students favor having a student lounge.

Example

A customer calculated that 0.294 people in a grocery store buy green beans. What percent of people in the store buy green beans?

STEP 1 **Multiply the decimal by 100.** This moves the decimal point *2* places to the *right*.

$0.294 \times 100 = 0.29.4 \longrightarrow 29.4$ Drop zeros that are not needed.

STEP 2 Write the percent sign.

$0.294 = 29.4\%$

(ON YOUR OWN)

According to a survey, 0.8 of the students in Neela's school like the idea of extending the school day to 5:00 PM. What percent of students want to extend the school day?

Practice

Building Skills

Write each decimal as a percent.

1. 0.06

2. 0.64

3. 0.052

4. 0.888

5. 0.7

6. 0.005

7. 0.0024

8. 0.908

9. 1.4

10. 1.063

Problem Solving

Solve.

11. In Serena's class, 0.45 of the students do not want to go to school through July. What is 0.45 written as a percent?

12. Your friend estimates that about 0.95 of the students in his high school would like to order take-out food for lunch. What percent of the students is that?

13. Out of all the members of the school drama club, only 0.35 can sing, dance, and act. What percent of the drama club can sing, dance, and act?

14. In a survey of high-school seniors, 0.05 said that they would like to spend another year at school. What percent of the seniors surveyed would like to stay in high school for another year?

15. In a survey that Golda took, 0.652 of the people she surveyed said that students should take a class about money. What percent of the people that Golda surveyed *did not* think that students should take a class about money?

16. Of the 400 students in Moira's class, 0.98 of them say that they plan to become teachers. What percent is this?

67

Name _____ Date _____

LESSON 30 — Converting Percents to Decimals

Sometimes, to solve a problem it is easier to work with percents that have been changed to decimals. You can write any percent as a decimal.

- Get rid of the percent sign. Add a decimal point if needed. A decimal point to the right of a whole number does not change its value.

- Move the decimal point *2* places to the *left*. Sometimes you may need to add a zero as a placeholder.

Why did you move the decimal? When you wrote decimals as percents, you moved the decimal point 2 places to the right. Now, when you change percents to decimals, you will move the decimal point in the opposite direction: 2 places to the left. Remember, *percent* means *per hundred* or *divided by one hundred*.

Example

Only 6% of teens questioned said that watching movies at home is better than seeing them in the theater. What is 6% written as a decimal?

STEP 1 Remove the percent sign.
6% = 6.

STEP 2 Move the decimal point two places to the left.

6. ⟶ 0.06.

It is also correct to write a 0 to the left of the decimal point in the ones place.

6% written as a decimal is 0.06.

ON YOUR OWN

The newspaper reported that 25.5% of people surveyed watched the news every night. How would you write this as a decimal?

Practice

Move the decimal point 2 places to the left.

Building Skills

Write each percent as a decimal.

1. 85%   **2.** 65%   **3.** 12%   **4.** 48%

5. 73%   **6.** 29%   **7.** 4%   **8.** 34.7%

9. 150%   **10.** 2.5%   **11.** 3.1%   **12.** 0.3%

Problem Solving

Solve.

13. According to a survey, 74% of teens believe that the prices for a movie are too high. How would you write that percent as a decimal?

14. Moviegoers agree that prices are 40% too high. What decimal is equal to this percent?

15. Teens were asked about their favorite kinds of movies. Forty-eight percent said that comedies were their favorite. What is the decimal equivalent?

16. Of teens surveyed, only 5% said that movie theaters charge fair prices for popcorn and candy. Write this percent as a decimal.

17. A movie critic has given a thumbs-up to 62.5% of the movies she has rated. Write a decimal equal to the percent of movies that she did *not* like.

18. A young filmmaker says that she always gives a 120% effort on every film she shoots. What decimal is equal to 120%?

LESSON 31 Find the Part

When you are solving percent problems, you're looking for a missing piece. The percent triangle helps you to find that piece. A percent triangle shows how the three pieces are related.

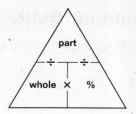

Look at the triangle. To find the part, cover the word *part*. The remaining pieces are connected by a multiplication sign. Multiply the pieces you have to find the part.

Example

Find 25% of 75.

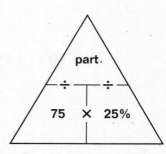

STEP 1 **Identify the pieces you have.**
25% is the percent. 75 is the whole.

STEP 2 **Write a percent sentence.**
part = whole × percent

STEP 3 **Replace the words with numbers.**
part = 75 × 25%

STEP 4 **Multiply to find the answer.**
Remember to convert your percent to a decimal.
75 × 25% = 75 × 0.25 = 18.75.

25% of 75 is 18.75.

ON YOUR OWN

There are 60 students in Mr. Watt's math class. 65% of them have jobs after school. How many students have jobs after school?

Practice

Building Skills

Find the part.

1. 82% of 50

2. 31% of 66

3. 10% of 45

4. 50% of 600

5. 23% of 90

6. 2.5% of 22

7. 60% of 60

8. 64.7% of 12

9. 150% of 4

Problem Solving

Solve.

10. Twenty-eight percent of the 150 students in the senior class are either on a team or in a club. How many seniors are either on a team or in a club?

11. Seventy-five percent of the actors in a play are on stage for the first time. There are 32 actors in the play. How many are on stage for the first time?

12. There are 40 stores in the mall. Of these stores, 62.5% sell clothing. How many stores in the mall sell clothing?

13. In one baseball game, 37.5% of the balls hit were fly balls. If 24 balls were hit in that game, how many were fly balls?

14. You are collecting for a local charity. Seventy percent of the 140 houses that you called on made a promise, or pledge, to give money. How many houses made a pledge?

15. A company's sales goal for March was $250,000. The company reached 80% of that goal. What was the amount of their sales in March?

LESSON 32 Find the Whole

Sixteen teens have after-school jobs at a local video store. This is 20% of the students who applied for jobs. How many teens applied for jobs at the video store?

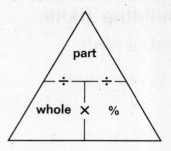

You used the percent triangle to help you find the part. Sometimes you will be asked to find the whole. The percent triangle can help you find the whole, too.

To find the whole, cover the word *whole*. The remaining pieces are connected with a division sign. Divide the part by the percent to find the whole. Remember to rewrite the percent as a decimal before you divide.

Example

30 is 60% of what number?

STEP 1 Identify the pieces you have.
30 is the part. 60% is the percent.

STEP 2 Write a percent sentence.
whole = part ÷ percent

STEP 3 Replace the words with numbers.
whole = 30 ÷ 60%

STEP 4 Divide to find the answer.
30 ÷ 60% = 30 ÷ 0.60 = 50

30 is 60% of 50.

(ON YOUR OWN)

12 is 24% of what number?

Practice

Rewrite the percent as a decimal.

Building Skills

Find the whole.

1. Forty-two is 50% of what number?

2. Forty-eight is 40% of what number?

3. Ninety is 30% of what number?

4. Fifteen is 125% of what number?

Find *n*. (Think of *n* as *what number*.)

5. Twenty-five percent of *n* is 20.

6. One hundred twenty percent of *n* is 18.

7. One hundred twenty is 125% of *n*.

8. Twenty-five percent of *n* is 72.

Problem Solving

Solve.

9. Twenty-four college students were hired as ushers for a concert. This is 30% of the students who applied for the job. How many college students applied to be ushers?

10. Thirty-two businesses in town receive an award for supporting the community. This is 16% of all of the businesses in town. How many businesses are there in town?

11. Six percent of those who applied were accepted into the summer-study program at the college. Three hundred were accepted. How many applied?

12. Three hundred thirty deer were counted by park rangers. This is 165% of the number expected. How many deer did park rangers expect to count?

13. Forty-five parents, which is 37.5% of all the parents, came to see the school play. How many parents are there in all?

14. Seventy-five of the students who entered the art contest won prizes. This is $12\frac{1}{2}$% of all the students who entered. How many students entered the contest?

Name _____   Date _____

LESSON 33   Find the Percent

Two hundred radio listeners called in to name their favorite group. Forty named the same group as their favorite. What percent of those who called in named this group?

You have used the percent triangle to find the part and to find the whole. Now you can use the percent triangle to find the percent.

To find the percent, cover the symbol for *percent*. The remaining pieces are connected with a division sign. Divide the part by the whole to find the percent. The answer is a decimal. Multiply the decimal by 100 to find the percent form.

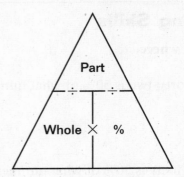

Example

33 is what percent of 150?

STEP 1 Identify the pieces you have.
33 is the part. 150 is the whole.

STEP 2 Write a percent sentence.
percent = (part ÷ whole) × 100

STEP 3 Replace the words with numbers.
percent = (33 ÷ 150) × 100

STEP 4 Divide. Then multiply by 100 and add the % sign.
(33 ÷ 150) × 100 = 0.22 × 100 = 22%

33 is 22% of 150.

(ON YOUR OWN)

What percent of 88 is 22?

Fractions, Ratios, and Percents, SV 0436-0

Practice

Building Skills

Find each percent.

1. What percent of 240 is 60?

2. What percent of 240 is 90?

3. What percent of 288 is 72?

4. What percent of 200 is 140?

5. What percent of 360 is 18?

6. Thirteen is what percent of 80?

7. Twenty-seven is what percent of 50?

8. Six is what percent of 40?

9. Forty is what percent of 25?

10. Twelve is what percent of 75?

Problem Solving

Solve.

11. Only 400 of 1,200 teens chose country music as their favorite type of music. What percent of the teens chose country music?

12. A disc jockey plays 48 different songs at a party. Thirty-six of those songs are fast songs. What percent of the songs played are fast songs?

13. Thirty-two students are invited to go sightseeing. However, 144 students show up! Describe the number who came to go sightseeing as a percent of those who were invited. (*Hint:* The answer is more than 100%.)

14. Fifty-six of 224 passengers on an airplane slept on the flight. What percent of the passengers got some sleep?

LESSON 34   What Is a Percent Equation?

You have worked with different equations to find the missing piece in a percent problem. There is another way to solve percent problems. Use a **percent equation** to solve any type of percent problem. Write the equation as simply as you can. Use the **variable,** *n*, to stand for the missing number.

The rules on the right will help you.

> *Rules for Reading and Writing a Percent Equation*
>
>
>
> 1. Restate the question as simply as possible. Replace any percent in the problem with a decimal.
>
> 2. Write the equation the way you read the problem. Use *n* for the missing number.

Example

30 whales were spotted offshore. This is 60% of the total expected. How many whales were expected?

STEP 1 Restate the problem as simply as possible. Replace any percent with a decimal.
30 is 0.6 of what number?

STEP 2 Write the equation the way you read the problem. Use *n* for the missing number.
$30 = 0.6 \times n$

30	is	0.6	of	what number?
30	=	0.6	×	*n*

STEP 3 Solve for *n*.
$$30 = 0.6 \times n$$
$$\frac{30}{0.6} = \frac{0.6}{0.6} \times n$$
$$50 = n$$

50 whales were expected.

(ON YOUR OWN)

Last year Ms. Salas had 30 students in her class. There were 120 students in the school. What percent of the students were in Ms. Salas's class?

Practice

Building Skills

Write a percent equation for each problem.

1. What percent of 60 is 24?

2. What is 32% of 80?

3. What percent of 64 is 12?

4. Eighty percent of what number is 16?

5. What percent of 240 is 180?

6. What is 28% of 54?

7. Sixty-two percent of what number is 16?

8. Sixty is what percent of 20?

9. What is 2% of 10?

10. Ten percent of what number is 12?

11. What percent of 3.6 is 1.8?

12. What is 2.2% of 200?

13. Forty-four percent of what number is 22?

14. Sixteen is what percent of 68?

15. What is 2.32% of 50?

16. Seventy-five percent of what number is 33?

77

LESSON 35 Creating a Percent Equation from a Word Problem

A word problem can be solved using a percent equation, too. First, cut out all the details when you rewrite the problem. Then, use the numbers that you have. Usually you are given two numbers and asked to find the third number. You can replace the word *percent* with the word *decimal*. This reminds you that your answer, *n*, will be the decimal form of the percent.

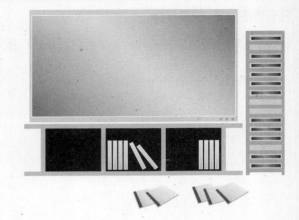

Example

Maureen has a collection of 80 DVDs. She loaned 16 of them to her friends. What percent of the DVDs did she loan?

STEP 1 Restate the question as simply as possible. Replace the word *percent* with the word *decimal*.
What percent of 80 is 16?
What decimal of 80 is 16?

STEP 2 Write the equation the way you read the problem. Use *n* for the missing number and solve.
$n \times 80 = 16$
$n = 0.2 = 0.2 \times 100 = 20\%$

Maureen loaned 20% of her DVD collection.

ON YOUR OWN

In a parade, 1,170 marchers each carry a musical instrument. This is 65% of the total marchers. How many marchers are in the parade?

Practice

Building Skills

Write a percent equation and solve each problem.

1. The mail carrier delivers the mail to 44 of the houses on Warren Street. This is 20% of all the houses on Warren Street. How many houses are on Warren Street?

2. Latisha saves 30% of her monthly take-home pay to buy a set of drums. Her monthly take-home pay is $1,340. How much does she save each month?

3. A professional volleyball team won 28 games and lost 22. What percent of their games did they lose?

4. Enrico sold 55 postcards from his collection. This was 11% of his collection. How many postcards did he have before he sold any?

5. A city recycles 85% of the newspapers sold there. The total amount of newspapers sold comes to 8 tons. How many tons get recycled?

6. Leanna spent $50 to buy hats with her school's name on them. She sold them all for a total of $175. By what percent did her money increase?

7. Eighteen students in the school system are new to the United States. This is 0.2% of all the students. How many students are in the school system?

8. In a beach volleyball play-off, 28% of the players are in a play-off for the first time. There are 150 players. How many are in their first play-off?

9. In one college, 386 of the 623 graduating seniors had jobs lined up after graduation. What percent of the seniors had jobs lined up? (Round your answer to the nearest whole percent.)

10. In one study 94.5%, or 821, of the people who do *not* exercise regularly said that they do not feel their best. How many total people were in the study? (Round your answer to the nearest whole number.)

LESSON 36 Percent Change

When you find a percent change, you are comparing a number and the amount that the number changes. Number amounts can increase (become larger) or decrease (become smaller). Using percents makes the comparison easier.

According to the newspaper headline, which company lost more workers?

NEWS
Monarch Supermarkets fires 12% of 750 employees. Ninety workers are out of a job.

NEWS
Powell Industries lets 792 of 3600 workers go – a 22% reduction.

Example

A company that had 30 workers now has only 24. What is the percent of decrease?

STEP 1 Subtract to find the amount of decrease or increase.
original − new
$30 − 24 = 6$
The amount of decrease is 6.

STEP 2 Write a percent equation and solve for n.
Remember to use n for the number you are looking for.
6 is what percent of 30?
$$6 = n \times 30$$
$$\frac{6}{30} = n \times \frac{30}{30}$$
$$0.2 = n$$

STEP 3 Change the decimal to a percent.
$0.2 \times 100\% = 20\%$

The percent of decrease is 20%.

ON YOUR OWN

There were 40 stores in a mall. Now there are 35. What is the percent of decrease?

Practice

Building Skills

Find the percent of increase or decrease.

Subtract to find the amount of decrease or increase. Then find the percent.

1. 60 teachers last year; 57 this year.

2. Original price was $60; new price is $45.

3. Original number of seats was 800; new number of seats is 850.

4. Original number of acres was 32; new number of acres is 16.

5. Old number was 40; new number is 48.

6. Last year's score was 50; this year's score is 42.

7. Original salary was $120; new salary is $100.

8. Original price was $85; new price is $76.50.

Problem Solving

Solve.

9. Antwon used to work 50 hours each month. Now he works 45 hours a month. What is the percent of decrease?

10. A surfboard factory has 32 employees. As sales go up, the number of workers increases to 40. What is the percent of increase?

11. A trainee's salary is raised from $500 per week to $550 per week. What is the percent of increase?

12. A group of music stores had 300 people working for them. Now there are only 220. What is the percent of decrease?

13. The round-trip fare from New York to San Francisco was $440. It has been reduced to $396. What is the percent of decrease?

14. Last season, Jamaal's batting average was 0.240. With all his injuries this year, his average dropped to 0.192. What is the percent of decrease?

LESSON 37 — Markup and Discount

To make money, stores charge you more for their items than they have to pay for them. The **markup** is the difference between what a store pays for the item and the price you pay for it.

You find the *percent of markup* the same way you find the *percent of increase*.

> markup = selling price to you − original cost to store

A **discount** is a cut in price.
You find the *percent of discount* the same way you find the *percent of decrease*.

> discount = original price − reduced price

Example

A store buys cell phones for $24 each and sells them for $36 each. What is the percent of markup?

STEP 1 Subtract to find the amount of increase or decrease.
selling price − original cost
$$36 - 24 = 12$$
The amount of increase is 12.

STEP 2 Write a percent equation and solve for *n*.
12 is what decimal of 24?
$$12 = n \times 24$$
$$\frac{12}{24} = \frac{24n}{24}$$
$$0.5 = n$$

STEP 3 Change the decimal to a percent.
$$0.5 \times 100\% = 50\%$$

The markup is 50%.

ON YOUR OWN

An $800 stereo system now sells for $750. What is the percent of discount?

Practice

Building Skills

Find the percent of markup or discount.

1. A store's original price for a camera was $285; the selling price is $240.

2. The original price was $600; discounted price is $480.

3. The original price was $84; discounted price is $75.60.

4. The store's cost is $60; selling price is $75.

5. The store's cost is $20; selling price is $26.

6. The original price was $300; discounted price is $210.

7. The store's cost is $25; selling price is $45.

8. The store's cost is $24; selling price is $28.

Problem Solving

Solve.

9. A $75 shirt is on sale for $60. What is the percent of discount?

10. A store paid $65 for a CD player that it sells for $80. What is the percent of markup?

11. A jacket used to sell for $160. It is now on sale for $128. What is the percent of discount?

12. A suit that sells for $300 has a markup of $120. What is the percent of markup?

13. Airfare from Miami to Phoenix usually costs $480. Now it is reduced to $410. What is the percent of discount?

14. The price for a new 32-inch television was $360. The new price is $280. What is the percent of discount?

LESSON 38   Simple Interest

When you borrow money from a bank, you pay interest on the loan. When you save money in a bank, you receive interest on the money you save.

To find simple interest, use this formula:

Interest = principal × rate × time

$$I = p \times r \times t$$

- **Interest** (I) is the money paid for using or saving other money.
- The **principal** (p) is the amount of money you start with.
- The **rate** (r) is the interest rate. This is the percent that the bank pays on the principal.
- The **time** (t) is the amount of time the money is in the bank.

Before you solve for interest, change the rate to a decimal and be sure the time is in years.

Example

How much simple interest will you earn when you put $600 in your bank at 4% interest for 6 months? How much money will be in this account at the end of 6 months?

STEP 1 Use the simple interest formula.
$I = p \times r \times t$
Substitute the numbers for the letters.
$I = p \times r \times t$
$I = \$600 \times 4\% \times 6$ months

STEP 2 Multiply to find the *simple interest.*
Change percents to decimals. Change months to years.
$I = \$600 \times 0.04 \times 0.5$
$I = \$12$
You will earn $12 in simple interest.

STEP 3 Add to find the *total amount.*
$\$600 + \$12 = \$612$
  ↓   ↓   ↓
Principal Interest Total amount

You will have $612 in your account after 6 months.

(ON YOUR OWN)

You save $3,000 for 2 years in a bank that pays an interest rate of 5% per year. Find the interest. Find the total amount in your account after 2 years.

Practice

Building Skills

Find the simple interest only.

1. $800 deposited at 3% for 2 years

2. $4,000 deposited at 4% for 1 year

3. $500 deposited at 5% for 2 years

4. $700 deposited at 3.5% for 0.5 year

Find the simple interest and total amount.

5. $1,000 deposited at 8% for 2 years

6. $900 deposited at 7.5% for 5 years

7. $1,300 deposited at 5.4% for 1 year

8. $800 deposited at 9% for 9 months

Problem Solving

Solve.

9. What will the total amount be after 3 years for a savings deposit of $320 at 6% interest?

10. A deposit is made of $900 at 5% interest for 6 months. How much interest will be earned during that time?

11. You deposit $1,100 for 9 months at 9% interest. How much interest will be earned during that time? How much money will be in the account at the end of this time?

12. Elena deposits $240 at 6% interest for $1\frac{1}{2}$ years. How much interest will be in her account at the end of that time?

13. You have $450 to put in a savings account. In 9 months, how much more will you have at $8\frac{1}{2}$% interest than at 8% interest? (Hint: You are finding the difference between two rates.)

14. Ming saved $1,000 at $6\frac{3}{4}$% for $2\frac{1}{2}$ years. How much money will be in his account at the end of $2\frac{1}{2}$ years?

LESSON 39 Compound Interest

You have learned that simple interest is paid only on the principal (the amount of money you start with). **Compound interest** is paid on the principal plus *any interest* that has built up in your account.

In other words, the interest you earn or pay gets interest, too.

You can use the following formula to find the balance in an account that earns compound interest.

$B = p (1 + r)^t$

The **balance** (**B**) of an account is the principal plus any interest earned.

p is principal.
r is rate of interest.
t is time. In this formula, t is an exponent.

> **Order of Operations**
> **P**arentheses
> **E**xponents
> **M**ultiplication or
> **D**ivision
> **A**ddition or
> **S**ubtraction

Example

You put $3,000 in a savings account that pays 4% interest compounded annually (each year). What is your balance after three years?

STEP 1 Write the formula for compound interest.

$$B = p(1 + r)^t$$

STEP 2 Substitute numbers into the formula.
Change the interest rate (r) to a decimal.

$$B = p(1 + r)^t$$
$$B = 3,000(1 + 0.04)^3$$

STEP 3 Solve using the order of operations. Note: In this problem, you will cube the number in parentheses, or multiply it by itself 3 times.
$B = 3,000(1.04)^3$
$B = 3,000(1.124864)$
$B = 3,374.592$

STEP 4 Round your answer to the nearest cent.

The balance in your account after 3 years is $3,374.59.

(**ON YOUR OWN**)

You put $400 in the bank for 4 years at a rate of 5% compounded annually. Find the total amount in your account after four years.

Practice

An exponent tells you how many times to multiply a number by itself.

Building Skills

Find the balance in each account using the compound interest formula.

1. $600 deposited at 4% after 2 years

2. $900 deposited at 6% after 3 years

3. $900 deposited at 4% after 5 years

4. $1,800 deposited at 6% after 4 years

5. $2,000 deposited at 4.5% after 1 year

6. $3,000 deposited at 4.25% after 12 years

7. $5,000 deposited at 4% after 5 years

8. $4,250 deposited at 6% after 4 years

Problem Solving

Solve.

9. Marshall put $500 in an account paying 8% compounded annually. How much was the account worth after 15 years?

10. A deposit of $300 earns 4% interest for 3 years compounded *quarterly* (every 3 months or 0.25 of a year). What will the total amount in this account be after three years? (*Hint:* $r = 0.01$, $t = 12$)

11. You deposit $1,000 at 10% compounded annually. What is your balance after a year and a half?

12. $9,000 is deposited in a savings account. The money is left there for two years. It earns 6% interest compounded semiannually (twice a year). How much money will be in the account after two years? (*Hint:* $r = 0.03$, $t = 4$)

13. Joshua's great-grandfather deposited $1,000 in an account that earns 5% interest compounded annually. The money stayed there for 100 years. What was the account worth at the end of 100 years?

14. An account pays 6% compounded monthly. What is the value of $150 left in the account for 1 year?

Glossary

balance (*B*) (page 86)
the principal plus any interest earned

compound interest (page 86)
the money paid based on the principal plus the interest already earned; the formula for finding the balance on an account that earns compound interest is $B = p(1 + r)^t$

cross multiply (page 52)
to multiply the numerator of one ratio by the denominator of the other ratio in a proportion

cross products (page 52)
the answer to multiplying the numerator of one ratio by the denominator of the other ratio in a proportion

decimal (page 24)
a number with one or more digits to the right of the decimal point

decimal point (page 24)
a symbol used to separate the ones and tenths places in a decimal

denominator (page 10)
the number below the fraction bar in a fraction

discount (page 82)
a cut in price

equation (page 76)
a statement of equality between two terms

equivalent (page 12)
equal or having the same value

equivalent fractions (page 12)
fractions that name the same amount

equivalent ratios (page 52)
ratios written as equivalent fractions

exponent (page 87)
a number that tells how many times a number, or base, is used as a factor

factor (page 46)
a whole number that evenly divides into another whole number

formula (page 84)
a rule that shows the relationship between two or more quantities; for example, the formula for calculating simple interest is $I = prt$

fraction (page 10)
a number that names part of a whole or part of a group

greatest common factor (GCF) (page 46)
the greatest number that is a factor of two or more numbers; for example, the greatest common factor of 18 and 30 is 6

improper fraction (page 18)
a fraction that has a numerator greater than or equal to the denominator; for example, $\frac{9}{5}$ or $\frac{37}{21}$

interest (*I*) (page 84)
the money paid for using or saving other money; the amount banks charge or pay on money borrowed or saved

interest rate (page 84)
the rate, given as a percent, used to determine interest; the percent a bank pays or charges on the principal

invert (page 42)
to turn upside down or reverse

least common denominator (LCD) (page 16)
the smallest number that the denominators of fractions being added or subtracted divide into evenly

least common multiple (LCM) (page 14)
the smallest number that is a multiple of two or more other numbers

markup (page 82)
the difference between what a store pays for an item and its selling price

mixed number (page 18)
a number that contains both a whole number and a fraction; for example, $3\frac{1}{2}$ or $12\frac{3}{5}$

numerator (page 10)
number above the fraction bar in a fraction

percent (page 60)
a special ratio that compares a number to 100; another way of saying per hundred

percent equation (page 76)
an equation used to show and solve percent problems

percent of change (increase or decrease) (page 80)
the percent a quantity increases or decreases from its original amount

principal (p) (page 84)
the amount of money you start with; for example, the original amount of a loan

proper fraction (page 20)
a fraction that has a numerator that is less than the denominator

proportion (page 52)
a statement that says two ratios are equal

rate (r) (page 48 and page 84)
a ratio that compares two quantities (amounts) measured in different units; also, the percent of interest paid on the principal

ratio (page 46)
compares two quantities, amounts, or numbers. Ratios must be written in simplest form.

rename (page 20)
change

simple interest (page 84)
money paid based only on the principal ($I = prt$)

simplest form (page 20)
a fraction in which 1 is the only number that divides evenly into the numerator and denominator

term (page 56)
the name given to each value in a proportion

time (t) (page 84)
the amount of time money is in the bank

unit price (page 50)
the unit rate that tells the price per unit; the cost of one item

unit rate (page 50)
the rate for one unit of a quantity. The denominator for any unit rate is 1.

variable (page 76)
a letter used to represent the number you are trying to find

Answer Key

PAGES 4–8

1. $\dfrac{5}{12}$

2. $\dfrac{24}{24}$

3. $\dfrac{4}{8}$

4. $\dfrac{10}{10}$

5. $\dfrac{6}{15} \div \dfrac{3}{3} = \dfrac{2}{5}$

6. $\dfrac{4}{6} \div \dfrac{2}{2} = \dfrac{2}{3}$

7. $\dfrac{6}{12} \div \dfrac{6}{6} = \dfrac{1}{2}$

8. $\dfrac{15}{18} \div \dfrac{3}{3} = \dfrac{5}{6}$

9. $\dfrac{6}{8} \div \dfrac{2}{2} = \dfrac{3}{4}$

10. $\dfrac{8}{16} \div \dfrac{8}{8} = \dfrac{1}{2}$

11. multiples of 3: 3, 6, 9, 12; multiples of 6: 6, 12, 18; therefore, the LCM is 6

12. multiples of 2: 2, 4, 6, 8; multiples of 6: 6, 12, 18; therefore, the LCM is 6

13. multiples of 4: 4, 8, 12, 16, 20, 24, 28; multiples of 7: 7, 14, 21, 28; therefore, the LCM is 28

14. multiples of 5: 5, 10, 15; multiples of 2: 2, 4, 6, 8, 10, 12; therefore, the LCM is 10

15. $\dfrac{1}{8} + \dfrac{3}{8} = \dfrac{4}{8} = \dfrac{1}{2}$

16. $\dfrac{1}{6} + \dfrac{1}{6} = \dfrac{2}{6} = \dfrac{1}{3}$

17. $\dfrac{2}{5} + \dfrac{3}{10} = \dfrac{4}{10} + \dfrac{3}{10} = \dfrac{7}{10}$

18. $3\dfrac{5}{8} + 2\dfrac{1}{2} = 3\dfrac{5}{8} + 2\dfrac{4}{8} = 5\dfrac{9}{8} = 6\dfrac{1}{8}$

19. $\dfrac{1}{3} + \dfrac{1}{6} = \dfrac{2}{6} + \dfrac{1}{6} = \dfrac{3}{6} = \dfrac{1}{2}$

20. $4\dfrac{5}{6} + 1\dfrac{1}{2} = 4\dfrac{5}{6} + 1\dfrac{3}{6} = 5\dfrac{8}{6} = 6\dfrac{2}{6} = 6\dfrac{1}{3}$

21. $\dfrac{5}{8} - \dfrac{3}{8} = \dfrac{2}{8} = \dfrac{1}{4}$

22. $\dfrac{5}{6} - \dfrac{1}{6} = \dfrac{4}{6} = \dfrac{2}{3}$

23. $\dfrac{1}{2} - \dfrac{1}{6} = \dfrac{3}{6} - \dfrac{1}{6} = \dfrac{2}{6} = \dfrac{1}{3}$

24. $\dfrac{3}{4} - \dfrac{1}{6} = \dfrac{9}{12} - \dfrac{2}{12} = \dfrac{7}{12}$

25. $6\dfrac{1}{3} - 2\dfrac{5}{6} = 6\dfrac{2}{6} - 2\dfrac{5}{6} = 5\dfrac{8}{6} - 2\dfrac{5}{6} = 3\dfrac{3}{6} = 3\dfrac{1}{2}$

26. $5\dfrac{1}{8} - 2\dfrac{3}{4} = 5\dfrac{1}{8} - 2\dfrac{6}{8} = 4\dfrac{9}{8} - 2\dfrac{6}{8} = 2\dfrac{3}{8}$

27. $\dfrac{3}{4} \times \dfrac{7}{8} = \dfrac{21}{32}$

28. $\dfrac{4}{5} \times \dfrac{5}{6} = \dfrac{20}{30} = \dfrac{2}{3}$

29. $\dfrac{9}{10} \times \dfrac{5}{6} = \dfrac{45}{60} = \dfrac{3}{4}$

30. $1\dfrac{1}{3} \times 1\dfrac{2}{3} = \dfrac{4}{3} \times \dfrac{5}{3} = \dfrac{20}{9} = 2\dfrac{2}{9}$

31. $3\dfrac{1}{8} \times 3\dfrac{1}{5} = \dfrac{25}{8} \times \dfrac{16}{5} = \dfrac{400}{40} = 10$

32. $\dfrac{8}{9} \times \dfrac{15}{16} = \dfrac{120}{144} = \dfrac{10}{12} = \dfrac{5}{6}$

33. $\dfrac{2}{3} \div \dfrac{1}{3} = \dfrac{2}{3} \times \dfrac{3}{1} = \dfrac{6}{3} = 2$

34. $\dfrac{5}{8} \div \dfrac{1}{4} = \dfrac{5}{8} \times \dfrac{4}{1} = \dfrac{20}{8} = \dfrac{5}{2} = 2\dfrac{1}{2}$

35. $2\dfrac{1}{2} \div 1\dfrac{1}{2} = \dfrac{5}{2} \div \dfrac{3}{2} = \dfrac{5}{2} \times \dfrac{2}{3} = \dfrac{10}{6} = \dfrac{5}{3} = 1\dfrac{2}{3}$

36. $2\dfrac{4}{5} \div 1\dfrac{1}{2} = \dfrac{14}{5} \div \dfrac{3}{2} = \dfrac{14}{5} \times \dfrac{2}{3} = \dfrac{28}{15} = 1\dfrac{13}{15}$

37. $\dfrac{4}{5} \div \dfrac{1}{5} = \dfrac{4}{5} \times \dfrac{5}{1} = \dfrac{20}{5} = 4$

38. $4\dfrac{1}{5} \div 2\dfrac{1}{3} = \dfrac{21}{5} \div \dfrac{7}{3} = \dfrac{21}{5} \times \dfrac{3}{7} = \dfrac{63}{35} = \dfrac{9}{5} = 1\dfrac{4}{5}$

39. $\dfrac{1}{2}$; 1:2; 1 to 2

40. $\dfrac{1}{3}$; 1:3; 1 to 3

41. $\dfrac{12 \text{ pages}}{3 \text{ minutes}} = \dfrac{12 \text{ pages} \div 3}{3 \text{ minutes} \div 3} = \dfrac{4 \text{ pages}}{1 \text{ minute}}$

42. $\dfrac{80 \text{ kilometers}}{4 \text{ hours}} = \dfrac{80 \text{ kilometers} \div 4}{4 \text{ hours} \div 4} = \dfrac{20 \text{ kilometers}}{1 \text{ hour}}$

43. $\dfrac{120 \text{ words}}{2 \text{ minutes}} = \dfrac{120 \text{ words} \div 2}{2 \text{ minutes} \div 2} = \dfrac{60 \text{ words}}{1 \text{ minute}}$

44. $\dfrac{60 \text{ miles}}{3 \text{ gallons}} = \dfrac{60 \text{ miles} \div 3}{3 \text{ gallons} \div 3} = \dfrac{20 \text{ miles}}{1 \text{ gallon}}$

45. $\dfrac{\$1.32}{6 \text{ cans}} = \dfrac{\$1.32 \div 6}{6 \text{ cans} \div 6} = \$0.22$; The rate is 22 cents per can.

46. $\dfrac{\$2.10}{3 \text{ pounds}} = \dfrac{\$2.10 \div 3}{3 \text{ pounds} \div 3} = \dfrac{\$0.70}{\text{pound}}$; The rate is 70 cents per pound.

47. $3 \times 10 = 30$; $6 \times 8 = 48$; $\dfrac{3}{8} \neq \dfrac{6}{10}$

48. $4 \times 27 = 108$; $9 \times 12 = 108$; $\dfrac{4}{9} = \dfrac{12}{27}$

49. $5 \times 10 = 50$; $8 \times 16 = 128$; $\dfrac{5}{8} \neq \dfrac{16}{10}$

50. $4 \times 21 = 84$; $7 \times 12 = 84$; $\dfrac{4}{7} = \dfrac{12}{21}$

51. $9 \times n = 6 \times 6$

$9n = 36$

$\dfrac{9n}{9} = \dfrac{36}{9}$

$n = 4$

52. $15 \times n = 20 \times 45$

$15n = 900$

$\dfrac{15n}{15} = \dfrac{900}{15}$

$n = 60$

53. $12 \times n = 4 \times 3$

$12n = 12$

$\dfrac{12n}{12} = \dfrac{12}{12}$

$n = 1$

54. $16 \times n = 8 \times 48$

$16n = 384$

$\dfrac{16n}{16} = \dfrac{384}{16}$

$n = 24$

55. $\dfrac{2}{9} = \dfrac{n}{225}$

$9 \times n = 2 \times 225$

$9n = 450$

$\dfrac{9n}{9} = \dfrac{450}{9}$

$n = 50$ students

56. $\dfrac{420}{35} = \dfrac{n}{1}$

$35 \times n = 1 \times 420$

$35n = 420$

$\dfrac{35n}{35} = \dfrac{420}{35}$

$n = \$12$ per hour

57. $0.065 = 6.5\%$

58. $\dfrac{7}{8} = 7 \div 8 = 0.875 = 87.5\%$

59. $\dfrac{5}{16} = 0.3125 = 31.25\%$

60. $0.062 = 6.2\%$

61. $32\% = \dfrac{32}{100} = \dfrac{32 \div 4}{100 \div 4} = \dfrac{8}{25}$

62. $125\% = \dfrac{125}{100} = \dfrac{125 \div 25}{100 \div 25} = \dfrac{5}{4} = 1\dfrac{1}{4}$

63. $42\% = \dfrac{42}{100} = \dfrac{42 \div 2}{100 \div 2} = \dfrac{21}{50}$

64. $115\% = \dfrac{115}{100} = \dfrac{115 \div 5}{100 \div 5} = \dfrac{23}{20} = 1\dfrac{3}{20}$

65. $22\% \times 60 = 0.22 \times 60 = 13.2$

66. $(24 \div 40) \times 100\% = 0.60 \times 100\% = 60\%$

67. $1.2 \div 40\% = 1.2 \div 0.40 = 3$

68. $(22 \div 88) \times 100\% = 0.25 \times 100\% = 25\%$

69. $30 \div 75\% = 30 \div 0.75 = 40$ DVDs

70. $100\% - 15\% = 85\%$

$40 \times 85\% = 40 \times 0.85 = \$34$

71. $I = \$600 \times 0.04 \times 2 = \$48$

72. $I = \$4,000 \times 0.055 \times 0.75 = \$165$

73. $B = 250(1 + 0.025)^2 = 250(1.025)^2 = 250(1.0506) = \$262.65$

74. $B = 100(1 + 0.075)^2 = 100(1.075)^2 = 100(1.1556) = \$115.56$

LESSON 1

ON YOUR OWN (PAGE 10): $\dfrac{3}{5}$
PAGE 11

1. There are 3 quarters out of 4 that make a dollar: $\dfrac{3}{4}$.

2. There are 7 total pieces of fruit, of which 3 are apples: $\dfrac{3}{7}$.

3. There are 2 quarts out of 4 that make a gallon: $\dfrac{1}{2}$.

4. $\dfrac{11}{18}$ 5. $\dfrac{11}{12}$ 6. $\dfrac{7}{12}$

7. $\dfrac{8}{24}$ or $\dfrac{1}{3}$ 8. $\dfrac{7}{12}$ 9. $\dfrac{19}{24}$

10. 9 out of 40 hours or $\dfrac{9}{40}$

11. 40 minutes out of 60 or $\dfrac{40}{60}$ or $\dfrac{2}{3}$

12. 12 pictures out of 16 or $\dfrac{12}{16}$ or $\dfrac{3}{4}$

13. 57 made out of 60 tries or $\dfrac{57}{60}$

LESSON 2

ON YOUR OWN (PAGE 12): $\dfrac{5}{6}$*
PAGE 13

1. $48 \div \dfrac{2}{2} = \dfrac{1}{2}$*   **2.** $\dfrac{10}{15} \div \dfrac{3}{3} = \dfrac{2}{3}$*

3. $\dfrac{6}{12} \div \dfrac{6}{6} = \dfrac{1}{2}$*   **4.** $\dfrac{18}{24} \div \dfrac{6}{6} = \dfrac{3}{4}$*

5. $\dfrac{1}{2} \times \dfrac{6}{6} = \dfrac{6}{12}$ **6.** $\dfrac{3}{5} \times \dfrac{2}{2} = \dfrac{6}{10}$

7. $\dfrac{2}{3} \times \dfrac{3}{3} = \dfrac{6}{9}$ **8.** $\dfrac{9}{10} \times \dfrac{3}{3} = \dfrac{27}{30}$

9. $\dfrac{6}{10} \div \dfrac{2}{2} = \dfrac{3}{5}$ **10.** $\dfrac{5}{8} \times \dfrac{2}{2} = \dfrac{10}{16}$

11. $\dfrac{30}{40} \div \dfrac{10}{10} = \dfrac{3}{4}$ **12.** $\dfrac{16}{20} \div \dfrac{4}{4} = \dfrac{4}{5}$*

13. $\dfrac{25}{50} \div \dfrac{25}{25} = \dfrac{1}{2}$*   **14.** $\dfrac{25}{30} \div \dfrac{5}{5} = \dfrac{5}{6}$

* other answers possible

LESSON 3

ON YOUR OWN (PAGE 14): 20 days
PAGE 15

1. multiples of 9: 9, 18, 27, 36; multiples of 12: 12, 24, 36; therefore, the LCM is 36.

2. multiples of 4: 4, 8, 12, 16, 20; multiples of 8: 8, 16, 24, 32; therefore, the LCM is 8.

3. multiples of 6: 6, 12, 18, 24, 30; multiples of 9: 9, 18, 27, 36; therefore, the LCM is 18.

4. multiples of 8: 8, 16, 24, 32, 40, 48, 56, 64, 72; multiples of 9: 9, 18, 27, 36, 45, 54, 63, 72; therefore, the LCM is 72.

5. multiples of 6: 6, 12, 18, 24, 30; multiples of 8: 8, 16, 24, 32; therefore, the LCM is 24.

6. multiples of 10: 10, 20, 30, 40; multiples of 15: 15, 30, 45; therefore, the LCM is 30.

7. multiples of 6: 6, 12, 18, 24, 30, 36, 42; multiples of 7: 7, 14, 21, 28, 35, 42; therefore, the LCM is 42.

8. multiples of 5: 5, 10, 15, 20, 25, 30, 35, 40; multiples of 8: 8, 16, 24, 32, 40; therefore, the LCM is 40.

9. multiples of 4: 4, 8, 12, 16, 20; multiples of 10: 10, 20, 30, 40; therefore, the LCM is 20.

10. multiples of 6: 6, 12, 18, 24; multiples of 4: 4, 8, 12, 16; therefore, the LCM is 12 days.

11. multiples of 7: 7, 14, 21, 28, 35; multiples of 5: 5, 10, 15, 20, 25, 30, 35; therefore, the LCM is 35 days.

12. multiples of 8: 8, 16, 24, 32, 40; multiples of 10: 10, 20, 30, 40; therefore, the LCM is 40 minutes.

13. multiples of 2: 2, 4, 6, 8, 10, 12, 14; multiples of 7: 7, 14, 21; therefore, the LCM is 14 days.

14. multiples of 2: 2, 4, 6, 8; multiples of 3: 3, 6, 9; therefore, the LCM is 6 days.

15. multiples of 3: 3, 6, 9, 12, 15; multiples of 5: 5, 10, 15, 20; therefore, the LCM is 15 days.

LESSON 4

ON YOUR OWN (PAGE 16): $\frac{1}{3} < \frac{2}{5}$; Brendan has completed more.

PAGE 17

1. $\frac{5}{6} \times \frac{2}{2} = \frac{10}{12}$
$\frac{1}{4} \times \frac{3}{3} = \frac{3}{12}$
$\frac{10}{12} > \frac{3}{12}; \frac{5}{6} > \frac{1}{4}$

2. $\frac{1}{5} \times \frac{2}{2} = \frac{2}{10}$
$\frac{2}{10} = \frac{2}{10}; \frac{1}{5} = \frac{2}{10}$

3. $\frac{3}{4} \times \frac{2}{2} = \frac{6}{8}$
$\frac{6}{8} < \frac{7}{8}; \frac{3}{4} < \frac{7}{8}$

4. $\frac{4}{5} \times \frac{3}{3} = \frac{12}{15}$
$\frac{2}{3} \times \frac{5}{5} = \frac{10}{15}$
$\frac{12}{15} > \frac{10}{15}; \frac{4}{5} > \frac{2}{3}$

5. $\frac{2}{5} \times \frac{2}{2} = \frac{4}{10}$
$\frac{1}{2} \times \frac{5}{5} = \frac{5}{10}$
$\frac{4}{10} < \frac{5}{10}; \frac{2}{5} < \frac{1}{2}$

6. $\frac{1}{4} \times \frac{2}{2} = \frac{2}{8}$
$\frac{3}{8} > \frac{2}{8}; \frac{3}{8} > \frac{1}{4}$

7. $\frac{9}{10} \times \frac{2}{2} = \frac{18}{20}$
$\frac{3}{4} \times \frac{5}{5} = \frac{15}{20}$
$\frac{18}{20} > \frac{15}{20}; \frac{9}{10} > \frac{3}{4}$

8. $\frac{3}{6} = \frac{6}{12}$
$\frac{6}{12} > \frac{4}{12}$
$\frac{3}{6} > \frac{4}{12}$

9. $\frac{5}{8} \times \frac{7}{7} = \frac{35}{56}$
$\frac{3}{7} \times \frac{8}{8} = \frac{24}{56}$
$\frac{35}{56} > \frac{24}{56}; \frac{5}{8} > \frac{3}{7}$

10. $\frac{2}{5} \times \frac{2}{2} = \frac{4}{10}$
$\frac{1}{2} \times \frac{5}{5} = \frac{5}{10}$
$\frac{4}{10} < \frac{5}{10} < \frac{7}{10};$
$\frac{2}{5} < \frac{1}{2} < \frac{7}{10}$

11. $\frac{2}{3} \times \frac{2}{2} = \frac{4}{6}$
$\frac{1}{2} \times \frac{3}{3} = \frac{3}{6}$
$\frac{3}{6} < \frac{4}{6} < \frac{5}{6}; \frac{1}{2} < \frac{2}{3} < \frac{5}{6}$

12. $\frac{1}{2} \times \frac{2}{2} = \frac{2}{4}$
$\frac{2}{4} < \frac{3}{4}; \frac{1}{2} < \frac{3}{4}$

Jerry stayed at the game longer.

13. $\frac{4}{5} \times \frac{2}{2} = \frac{8}{10}$
$\frac{1}{2} \times \frac{5}{5} = \frac{5}{10}$
$\frac{5}{10} < \frac{8}{10} < \frac{9}{10}; \frac{1}{2} < \frac{4}{5} < \frac{9}{10}$
Barbara traveled the farthest.

14. $\frac{2}{3} \times \frac{4}{4} = \frac{8}{12}$
$\frac{3}{4} \times \frac{3}{3} = \frac{9}{12}$
$\frac{8}{12} < \frac{9}{12}; \frac{2}{3} < \frac{3}{4}$
The first type of juice is less nutritious.

15. $\frac{1}{4} \times \frac{4}{4} = \frac{4}{16}$
$\frac{1}{8} \times \frac{2}{2} = \frac{2}{16}$
$\frac{2}{16} < \frac{3}{16} < \frac{4}{16}$
$\frac{1}{8} < \frac{3}{16} < \frac{1}{4}$
So, the sockets would be placed $\frac{1}{8}$ in., $\frac{3}{16}$ in., and $\frac{1}{4}$ in.

LESSON 5

ON YOUR OWN (PAGE 18): $\frac{29}{8}$

PAGE 19

1. $2\frac{5}{8} = \frac{21}{8}$

2. $3\frac{4}{5} = \frac{19}{5}$

3. $4\frac{5}{6} = \frac{29}{6}$

4. $2\frac{9}{10} = \frac{29}{10}$

5. $6\frac{2}{3} = \frac{20}{3}$

6. $4\frac{1}{2} = \frac{9}{2}$

7. $3\frac{2}{5} = \frac{17}{5}$

8. $4\frac{1}{8} = \frac{33}{8}$

9. $1\frac{7}{8} = \frac{15}{8}$

10. $2\frac{1}{4} = \frac{9}{4}$
$2 \times 4 = 8$
$8 + 1 = 9$
so, $2\frac{1}{4} = \frac{9}{4}$

11. $4\frac{2}{3} = \frac{14}{3}$
$3 \times 4 = 12$
$12 + 2 = 14$
so, $4\frac{2}{3} = \frac{14}{3}$

12. $1\frac{5}{8} = \frac{13}{8}$
$1 \times 8 = 8$
$8 + 5 = 13$
so, $1\frac{5}{8} = \frac{13}{8}$

13. $1\frac{1}{2} = \frac{3}{2}$
$1 \times 2 = 2$
$2 + 1 = 3$
so, $1\frac{1}{2} = \frac{3}{2}$

14. $2\frac{1}{4} = \frac{9}{4}$
$2 \times 4 = 8$
$8 + 1 = 9$
so, $2\frac{1}{4} = \frac{9}{4}$

15. $2\frac{3}{4} = \frac{11}{4}$
$2 \times 4 = 8$
$8 + 3 = 11$
so, $2\frac{3}{4} = \frac{11}{4}$

LESSON 6

ON YOUR OWN (PAGE 20): $1\frac{3}{5}$

PAGE 21

1. $\frac{17}{10} = 1\frac{7}{10}$

2. $\frac{23}{5} = 4\frac{3}{5}$

3. $\frac{18}{6} = 3$

4. $\frac{21}{4} = 5\frac{1}{4}$

5. $\frac{45}{8} = 5\frac{5}{8}$

6. $\frac{37}{7} = 5\frac{2}{7}$

7. $\frac{29}{6} = 4\frac{5}{6}$

8. $\frac{61}{8} = 7\frac{5}{8}$

9. $\frac{47}{5} = 9\frac{2}{5}$

10. $1\frac{1}{2}$ minutes

$90 \div 60 = 1\frac{30}{60} = 1\frac{1}{2}$

11. $1\frac{3}{4}$ hours

$\frac{7}{4} = 1\frac{3}{4}$

12. 5 quarts $= 1\frac{1}{4}$ gallons

$5 \div 4 = 1\frac{1}{4}$

13. 13 pairs of socks

$27 \div 2 = 13\frac{1}{2}$

14. $7\frac{1}{9}$ packages

$64 \div 9 = 7\frac{1}{9}$

15. $4\frac{1}{2}$ yards

$\frac{1}{2} \times 9 = \frac{9}{2}$

$\frac{9}{2} = 4\frac{1}{2}$

LESSON 7

ON YOUR OWN (PAGE 22): $3\frac{1}{4}$ yards
PAGE 23

1. $3\frac{12}{10} = 4\frac{1}{5}$

2. $2\frac{6}{8} = 2\frac{3}{4}$

3. $6\frac{4}{6} = 6\frac{2}{3}$

4. $5\frac{9}{6} = 6\frac{1}{2}$

5. $2\frac{5}{10} = 2\frac{1}{2}$

6. $4\frac{12}{16} = 4\frac{3}{4}$

7. $7\frac{9}{12} = 7\frac{3}{4}$

8. $3\frac{10}{8} = 4\frac{1}{4}$

9. $5\frac{8}{12} = 5\frac{2}{3}$

10. $2\frac{1}{2}$ cans

$\frac{4}{8} \div \frac{4}{4} = \frac{1}{2}$

so, $2\frac{4}{8} = 2\frac{1}{2}$

11. $9\frac{1}{5}$ gallons

$12 \div 10 = 1\frac{2}{10} = 1\frac{1}{5}$

$8 + 1 = 9$

$9 + \frac{1}{5} = 9\frac{1}{5}$

so, $8\frac{12}{10} = 9\frac{1}{5}$

12. $30\frac{3}{5}$ seconds

$\frac{6}{10} \div \frac{2}{2} = \frac{3}{5}$

so, $30\frac{6}{10} = 30\frac{3}{5}$

13. $7\frac{4}{5}$ packages

$\frac{20}{25} \div \frac{5}{5} = \frac{4}{5}$

so, $7\frac{20}{25} = 7\frac{4}{5}$

14. $5\frac{3}{4}$ yards

$\frac{12}{16} \div \frac{4}{4} = \frac{3}{4}$

so, $5\frac{12}{16} = 5\frac{3}{4}$

15. $8\frac{1}{5}$ boxes

$\frac{12}{10} = 1\frac{2}{10} = 1\frac{1}{5}$

$7 + 1\frac{1}{5} = 8\frac{1}{5}$

so, $7\frac{12}{10} = 8\frac{1}{5}$

LESSON 8

ON YOUR OWN (PAGE 24): 0.625
PAGE 25

1. $\frac{1}{2} = 0.5$

2. $\frac{3}{8} = 0.375$

3. $\frac{4}{5} = 0.8$

4. $0.95 = \frac{95}{100} = \frac{19}{20}$

5. $0.125 = \frac{1}{8}$

6. $0.72 = \frac{72}{100} = \frac{36}{50} = \frac{18}{25}$

7. $\frac{2}{5} = 0.4$

8. $\frac{7}{25} = 0.28$

9. $0.625 = \frac{5}{8}$

10. 0.75

3 quarters is less

than a dollar.

A quarter $= 0.25$

$3 \times 25 = 75$

$\frac{75}{100} = \frac{3}{4}$

so, $\frac{3}{4} = 0.75$

11. 0.875

$\frac{7}{8} = 7 \div 8$

$7 \div 8 = 0.875$

so, $\frac{7}{8} = 0.875$

12. 0.25

$\frac{1}{4} = 1 \div 4$

$1 \div 4 = 0.25$

so, $\frac{1}{4} = 0.25$

13. 0.625

$\frac{5}{8} = 5 \div 8$

$5 \div 8 = 0.625$

so, $\frac{5}{8} = 0.625$

14. $\frac{17}{20}$

$0.85 = \frac{85}{100}$

$\frac{85}{100} \div \frac{5}{5} = \frac{17}{20}$

so, $0.85 = \frac{17}{20}$

15. $\frac{5}{8}$

$0.625 = \frac{625}{1000}$

$\frac{625}{1000} \div \frac{25}{25} = \frac{25}{40}$

$\frac{25}{40} \div \frac{5}{5} = \frac{5}{8}$

so, $0.625 = \frac{5}{8}$

LESSON 9

ON YOUR OWN (PAGE 26): $1\frac{1}{2}$ hours
PAGE 27

1. $\frac{3}{4}$

$\frac{1}{8} + \frac{5}{8} = \frac{6}{8}$

$\frac{6}{8} \div \frac{2}{2} = \frac{3}{4}$

so, $\frac{1}{8} + \frac{5}{8} = \frac{3}{4}$

2. $\frac{2}{7}$

$\frac{1}{7} + \frac{1}{7} = \frac{2}{7}$

3. $\frac{1}{2}$

$\frac{3}{20} + \frac{7}{20} = \frac{10}{20}$

$\frac{10}{20} \div \frac{2}{2} = \frac{1}{2}$

so, $\frac{3}{20} + \frac{7}{20} = \frac{1}{2}$

4. $\frac{1}{2}$

$\frac{1}{4} + \frac{1}{4} = \frac{2}{4}$

$\frac{2}{4} \div \frac{2}{2} = \frac{1}{2}$

so, $\frac{1}{4} + \frac{1}{4} = \frac{1}{2}$

5. $\frac{4}{5}$

$\frac{1}{10} + \frac{7}{10} = \frac{8}{10}$

$\frac{8}{10} \div \frac{2}{2} = \frac{4}{5}$

so, $\frac{1}{10} + \frac{7}{10} = \frac{4}{5}$

6. $\frac{4}{5}$

$\frac{4}{15} + \frac{8}{15} = \frac{12}{15}$

$\frac{12}{15} \div \frac{3}{3} = \frac{4}{5}$

so, $\frac{4}{15} + \frac{8}{15} = \frac{4}{5}$

7. $1\frac{1}{2}$

$\frac{7}{12} + \frac{11}{12} = \frac{18}{12}$

$\frac{18}{12} = 1\frac{6}{12}$

$\frac{6}{12} \div \frac{2}{2} = \frac{1}{2}$

so, $\frac{7}{12} + \frac{11}{12} = 1\frac{1}{2}$

8. 1

$\frac{4}{9} + \frac{5}{9} = \frac{9}{9}$

$\frac{9}{9} = 1$

so, $\frac{4}{9} + \frac{5}{9} = 1$

Answer Key

Fractions, Ratios, and Percents, SV 0436-0

9. $1\frac{1}{4}$

$\frac{13}{16} + \frac{7}{16} = \frac{20}{16}$

$\frac{20}{16} = 1\frac{4}{16}$

$\frac{4}{16} \div \frac{4}{4} = \frac{1}{4}$

so, $\frac{13}{16} + \frac{7}{16} = 1\frac{1}{4}$

10. $\frac{4}{5}$ of the calls

$\frac{1}{5} + \frac{3}{5} = \frac{4}{5}$

11. $\frac{4}{5}$ of the book

$\frac{2}{5} + \frac{2}{5} = \frac{4}{5}$

12. $\frac{2}{3}$ of her clothing

$\frac{3}{6} + \frac{1}{6} = \frac{4}{6}$

$\frac{4}{6} \div \frac{2}{2} = \frac{2}{3}$

so, $\frac{3}{6} + \frac{1}{6} = \frac{2}{3}$

13. $\frac{1}{2}$ of the class

$\frac{3}{8} + \frac{1}{8} = \frac{4}{8}$

$\frac{4}{8} \div \frac{4}{4} = \frac{1}{2}$

so, $\frac{3}{8} + \frac{1}{8} = \frac{1}{2}$

14. $\frac{4}{5}$ of the total grade

$\frac{3}{5} + \frac{1}{5} = \frac{4}{5}$

15. $\frac{1}{2}$ of the students

$\frac{3}{10} + \frac{2}{10} = \frac{5}{10}$

$\frac{5}{10} \div \frac{2}{2} = \frac{1}{2}$

so, $\frac{3}{10} + \frac{2}{10} = \frac{1}{2}$

LESSON 10

ON YOUR OWN (PAGE 28): $1\frac{1}{6}$ hours

PAGE 29

1. $\frac{1}{4} + \frac{2}{3} = \frac{8}{12} + \frac{3}{12} = \frac{11}{12}$

2. $\frac{1}{2} + \frac{3}{5} = \frac{5}{10} + \frac{6}{10} = \frac{11}{10} = 1\frac{1}{10}$

3. $\frac{3}{8} + \frac{1}{6} = \frac{9}{24} + \frac{4}{24} = \frac{13}{24}$

4. $\frac{1}{9} + \frac{5}{6} = \frac{2}{18} + \frac{15}{18} = \frac{17}{18}$

5. $\frac{2}{3} + \frac{5}{6} = \frac{4}{6} + \frac{5}{6} = \frac{9}{6} = 1\frac{1}{2}$

6. $\frac{4}{5} + \frac{1}{2} = \frac{8}{10} + \frac{5}{10} = \frac{13}{10} = 1\frac{3}{10}$

7. $\frac{7}{8} + \frac{5}{6} = \frac{21}{24} + \frac{20}{24} = \frac{41}{24} = 1\frac{17}{24}$

8. $\frac{3}{4} + \frac{4}{5} = \frac{15}{20} + \frac{16}{20} = \frac{31}{20} = 1\frac{11}{20}$

9. $\frac{9}{10} + \frac{1}{4} = \frac{18}{20} + \frac{5}{20} = \frac{23}{20} = 1\frac{3}{20}$

10. $\frac{1}{2} + \frac{1}{4} = \frac{2}{4} + \frac{1}{4} = \frac{3}{4}$ of an hour

11. $\frac{3}{4} + \frac{5}{6} = \frac{18}{24} + \frac{20}{24} = \frac{38}{24} = 1\frac{14}{24} = 1\frac{7}{12}$ hours

12. $\frac{2}{8} + \frac{1}{3} = \frac{6}{24} + \frac{8}{24} = \frac{14}{24} = \frac{7}{12}$ of a disc

13. $\frac{5}{6} + \frac{2}{3} = \frac{5}{6} + \frac{4}{6} = \frac{9}{6} = 1\frac{1}{2}$ hours

14. $\frac{1}{3} + \frac{2}{7} = \frac{7}{21} + \frac{6}{21} = \frac{13}{21}$ of his earnings

15. $\frac{2}{8} + \frac{3}{6} = \frac{6}{24} + \frac{12}{24} = \frac{18}{24} = \frac{3}{4}$ of the day

LESSON 11

ON YOUR OWN (PAGE 30): $8\frac{1}{4}$ pounds

PAGE 31

1. $2\frac{1}{6} + 2\frac{2}{3} = 2\frac{1}{6} + 2\frac{4}{6} = 4\frac{5}{6}$

2. $5\frac{2}{5} + 2\frac{3}{10} = 5\frac{4}{10} + 2\frac{3}{10} = 7\frac{7}{10}$

3. $1\frac{1}{3} + 2\frac{1}{4} = 1\frac{4}{12} + 2\frac{3}{12} = 3\frac{7}{12}$

4. $3\frac{5}{8} + 2\frac{1}{6} = 3\frac{15}{24} + 2\frac{4}{24} = 5\frac{19}{24}$

5. $3\frac{1}{2} + 4\frac{2}{5} = 3\frac{5}{10} + 4\frac{4}{10} = 7\frac{9}{10}$

6. $6\frac{2}{3} + 2\frac{3}{5} = 6\frac{10}{15} + 2\frac{9}{15} = 8\frac{19}{15} = 9\frac{4}{15}$

7. $7\frac{4}{9} + 2\frac{1}{6} = 7\frac{8}{18} + 2\frac{3}{18} = 9\frac{11}{18}$

8. $2\frac{4}{5} + 3\frac{1}{6} = 2\frac{24}{30} + 3\frac{5}{30} = 5\frac{29}{30}$

9. $5\frac{7}{10} + 2\frac{3}{4} = 5\frac{14}{20} + 2\frac{15}{20} = 7\frac{29}{20} = 8\frac{9}{20}$

10. $4\frac{5}{8} + 2\frac{1}{2} = 4\frac{5}{8} + 2\frac{4}{8} = 6\frac{9}{8} = 7\frac{1}{8}$ miles

11. $1\frac{5}{6} + 1\frac{3}{4} = 1\frac{10}{12} + 1\frac{9}{12} = 2\frac{19}{12} = 3\frac{7}{12}$ hours

12. $15\frac{3}{5} + 17\frac{9}{10} = 15\frac{6}{10} + 17\frac{9}{10} = 32\frac{15}{10} = 33\frac{1}{2}$ minutes

13. $2\frac{1}{2} + 1\frac{3}{4} = 3\frac{5}{4} = 4\frac{1}{4}$ hours

14. $1\frac{2}{3} + 2\frac{1}{4} = 1\frac{8}{12} + 2\frac{3}{12} = 3\frac{11}{12}$ hours

15. $3\frac{2}{5} + 12\frac{4}{6} = 3\frac{12}{30} + 12\frac{20}{30} = 15\frac{32}{30} = 16\frac{2}{30} = 16\frac{1}{15}$ hours

LESSON 12

ON YOUR OWN (PAGE 32): $\frac{1}{2}$ pizza

PAGE 33

1. $\frac{4}{6} = \frac{2}{3}$

2. $\frac{2}{8} = \frac{1}{4}$

3. $\frac{3}{4}$

4. $\frac{2}{5}$

5. $\frac{3}{9} = \frac{1}{3}$

6. $\frac{9}{10}$

7. $\frac{4}{12} = \frac{1}{3}$

8. $\frac{8}{16} = \frac{1}{2}$

9. $\frac{7}{20}$

Answer Key
Fractions, Ratios, and Percents, SV 0436-0

10. $\frac{8}{10} - \frac{4}{10} = \frac{4}{10} = \frac{2}{5}$ of a mile

11. $\frac{12}{12} - \frac{1}{12} = \frac{11}{12}$ of the book

12. $\frac{5}{8} - \frac{3}{8} = \frac{2}{8} = \frac{1}{4}$ cup

13. $\frac{11}{12} - \frac{5}{12} = \frac{6}{12} = \frac{1}{2}$ of the garden

14. $\frac{8}{8} - \frac{1}{8} = \frac{7}{8}$ of the pizza

15. $\frac{7}{8} - \frac{5}{8} = \frac{2}{8} = \frac{1}{4}$ inch thicker

LESSON 13

ON YOUR OWN (PAGE 34): $\frac{1}{12}$ hour

PAGE 35

1. $\frac{7}{10} - \frac{5}{10} = \frac{2}{10} = \frac{1}{5}$

2. $\frac{2}{3} - \frac{1}{6} = \frac{4}{6} - \frac{1}{6} = \frac{3}{6} = \frac{1}{2}$

3. $\frac{4}{5} - \frac{1}{10} = \frac{8}{10} - \frac{1}{10} = \frac{7}{10}$

4. $\frac{1}{3} - \frac{1}{4} = \frac{4}{12} - \frac{3}{12} = \frac{1}{12}$

5. $\frac{5}{9} - \frac{1}{3} = \frac{5}{9} - \frac{3}{9} = \frac{2}{9}$

6. $\frac{7}{8} - \frac{3}{4} = \frac{7}{8} - \frac{6}{8} = \frac{1}{8}$

7. $\frac{11}{12} - \frac{3}{8} = \frac{22}{24} - \frac{9}{24} = \frac{13}{24}$

8. $\frac{15}{16} - \frac{3}{4} = \frac{15}{16} - \frac{12}{16} = \frac{3}{16}$

9. $\frac{5}{9} - \frac{1}{6} = \frac{10}{18} - \frac{3}{18} = \frac{7}{18}$

10. $\frac{9}{10} - \frac{1}{3} = \frac{27}{30} - \frac{10}{30} = \frac{17}{30}$ mile

11. $\frac{7}{10} - \frac{1}{5} = \frac{7}{10} - \frac{2}{10} = \frac{5}{10} = \frac{1}{2}$ mile

12. $\frac{7}{8} - \frac{5}{6} = \frac{21}{24} - \frac{20}{24} = \frac{1}{24}$ yard

13. $\frac{9}{10} - \frac{3}{4} = \frac{18}{20} - \frac{15}{20} = \frac{3}{20}$ mile

LESSON 14

ON YOUR OWN (PAGE 36): $\frac{3}{4}$ pound

PAGE 37

1. $3\frac{1}{3} - 1\frac{1}{2} = 3\frac{2}{6} - 1\frac{3}{6} = 2\frac{8}{6} - 1\frac{3}{6} = 1\frac{5}{6}$

2. $4\frac{5}{8} - 2\frac{1}{4} = 4\frac{5}{8} - 2\frac{2}{8} = 2\frac{3}{8}$

3. $5\frac{2}{3} - 1\frac{1}{2} = 5\frac{4}{6} - 1\frac{3}{6} = 4\frac{1}{6}$

4. $4\frac{3}{4} - 2\frac{2}{3} = 4\frac{9}{12} - 2\frac{8}{12} = 2\frac{1}{12}$

5. $3\frac{1}{4} - 1\frac{1}{2} = 3\frac{1}{4} - 1\frac{2}{4} = 2\frac{5}{4} - 1\frac{2}{4} = 1\frac{3}{4}$

6. $6\frac{3}{8} - 2\frac{7}{8} = 5\frac{11}{8} - 2\frac{7}{8} = 3\frac{4}{8} = 3\frac{1}{2}$

7. $8\frac{2}{5} - 1\frac{1}{3} = 8\frac{6}{15} - 1\frac{5}{15} = 7\frac{1}{15}$

8. $3\frac{1}{6} - 1\frac{11}{12} = 2\frac{14}{12} - 1\frac{11}{12} = 1\frac{3}{12} = 1\frac{1}{4}$

9. $12\frac{3}{8} - 6\frac{5}{6} = 12\frac{9}{24} - 6\frac{20}{24} = 11\frac{33}{24} - 6\frac{20}{24} = 5\frac{13}{24}$

10. $12\frac{2}{3} - 9\frac{1}{2} = 12\frac{4}{6} - 9\frac{3}{6} = 3\frac{1}{6}$ minutes

11. $2\frac{1}{3} - 1\frac{1}{4} = 2\frac{4}{12} - 1\frac{3}{12} = 1\frac{1}{12}$ pounds

12. $4\frac{1}{2} - 2\frac{3}{4} = 4\frac{2}{4} - 2\frac{3}{4} = 3\frac{6}{4} - 2\frac{3}{4} = 1\frac{3}{4}$ rolls

13. $2\frac{1}{3} - 1\frac{1}{4} = 2\frac{4}{12} - 1\frac{3}{12} = 1\frac{1}{12}$ bins

14. $3\frac{5}{8} - 1\frac{1}{4} = 3\frac{5}{8} - 1\frac{2}{8} = 2\frac{3}{8}$ lengths

15. $4\frac{5}{6} - 1\frac{3}{4} = 4\frac{10}{12} - 1\frac{9}{12} = 3\frac{1}{12}$ square feet

LESSON 15

ON YOUR OWN (PAGE 38): $\frac{1}{3}$ of the wall

PAGE 39

1. $\frac{5}{8} \times \frac{16}{25} = \frac{80}{200} = \frac{2}{5}$

2. $\frac{3}{4} \times \frac{1}{3} = \frac{3}{12} = \frac{1}{4}$

3. $\frac{2}{5} \times \frac{1}{6} = \frac{2}{30} = \frac{1}{15}$

4. $\frac{2}{9} \times \frac{3}{5} = \frac{6}{45} = \frac{2}{15}$

5. $\frac{5}{6} \times \frac{4}{5} = \frac{20}{30} = \frac{2}{3}$

6. $\frac{8}{9} \times \frac{3}{4} = \frac{24}{36} = \frac{2}{3}$

7. $\frac{15}{16} \times \frac{1}{10} = \frac{15}{160} = \frac{3}{32}$

8. $\frac{6}{7} \times \frac{14}{15} = \frac{84}{105} = \frac{4}{5}$

9. $\frac{9}{16} \times \frac{8}{15} = \frac{72}{240} = \frac{3}{10}$

10. $\frac{2}{3} \times \frac{1}{2} = \frac{2}{6} = \frac{1}{3}$ of the backyard

11. $\frac{1}{4} \times \frac{1}{2} = \frac{1}{8}$ of the watermelon

12. $\frac{4}{5} \times \frac{1}{4} = \frac{4}{20} = \frac{1}{5}$ cup

13. $\frac{9}{10} \times \frac{2}{3} = \frac{18}{30} = \frac{3}{5}$ mile

14. $1 \times \frac{2}{3} = \frac{2}{3}$ square yard

15. $\frac{1}{2} \times \frac{1}{3} = \frac{1}{6}$ cup

LESSON 16

ON YOUR OWN (PAGE 40): $117\frac{1}{2}$ square feet

PAGE 41

1. $3\frac{1}{3} \times 4\frac{1}{2} = \frac{10}{3} \times \frac{9}{2} = \frac{90}{6} = 15$

2. $1\frac{1}{4} \times 2\frac{2}{5} = \frac{5}{4} \times \frac{12}{5} = \frac{60}{20} = 3$

3. $3\frac{1}{8} \times 4\frac{4}{5} = \frac{25}{8} \times \frac{24}{5} = \frac{600}{40} = 15$

4. $5\frac{1}{4} \times 2\frac{4}{7} = \frac{21}{4} \times \frac{18}{7} = \frac{378}{28} = \frac{27}{2} = 13\frac{1}{2}$

5. $6\frac{1}{2} \times 3\frac{1}{5} = \frac{13}{2} \times \frac{16}{5} = \frac{208}{10} = \frac{104}{5} = 20\frac{4}{5}$

6. $3\frac{3}{4} \times 1\frac{1}{15} = \frac{15}{4} \times \frac{16}{15} = \frac{240}{60} = 4$

7. $4\frac{3}{8} \times 4\frac{4}{7} = \frac{35}{8} \times \frac{32}{7} = \frac{1120}{56} = 20$

8. $3\frac{2}{3} \times 1\frac{1}{11} = \frac{11}{3} \times \frac{12}{11} = \frac{132}{33} = 4$

9. $2\frac{13}{16} \times 4\frac{4}{9} = \frac{45}{16} \times \frac{40}{9} = \frac{1800}{144} = \frac{50}{4} = 12\frac{1}{2}$

10. $3\frac{1}{2} \times 5\frac{1}{3} = \frac{7}{2} \times \frac{16}{3} = \frac{112}{6} = \frac{56}{3} = 18\frac{2}{3}$ square feet

11. $2\frac{1}{2} \times 2\frac{1}{2} = \frac{5}{2} \times \frac{5}{2} = \frac{25}{4} = 6\frac{1}{4}$ cups

12. $5\frac{1}{2} \times 1\frac{1}{2} = \frac{11}{2} \times \frac{3}{2} = \frac{33}{4} = 8\frac{1}{4}$ laps

13. $6 \times 4\frac{1}{2} = \frac{6}{1} \times \frac{9}{2} = \frac{54}{2} = 27$ feet

14. $2\frac{3}{4} \times 4\frac{1}{3} = \frac{11}{4} \times \frac{13}{3} = \frac{143}{12} = 11\frac{11}{12}$ square inches

15. $7\frac{1}{3} \times 1\frac{1}{2} = \frac{22}{3} \times \frac{3}{2} = \frac{66}{6} = 11$ miles

LESSON 17

ON YOUR OWN (PAGE 42): 4 slices

PAGE 43

1. $\frac{5}{6} \div \frac{1}{3} = \frac{5}{6} \times \frac{3}{1} = \frac{15}{6} = 2\frac{3}{6} = 2\frac{1}{2}$

2. $\frac{3}{4} \div \frac{1}{2} = \frac{3}{4} \times \frac{2}{1} = \frac{6}{4} = \frac{3}{2} = 1\frac{1}{2}$

3. $\frac{9}{10} \div \frac{1}{4} = \frac{9}{10} \times \frac{4}{1} = \frac{36}{10} = 3\frac{6}{10} = 3\frac{3}{5}$

4. $\frac{7}{8} \div \frac{1}{3} = \frac{7}{8} \times \frac{3}{1} = \frac{21}{8} = 2\frac{5}{8}$

5. $\frac{4}{9} \div \frac{2}{3} = \frac{4}{9} \times \frac{3}{2} = \frac{12}{18} = \frac{2}{3}$

6. $\frac{5}{6} \div \frac{7}{8} = \frac{5}{6} \times \frac{8}{7} = \frac{40}{42} = \frac{20}{21}$

7. $\frac{15}{16} \div \frac{3}{16} = \frac{15}{16} \times \frac{16}{3} = \frac{240}{48} = 5$

8. $\frac{21}{25} \div \frac{7}{10} = \frac{21}{25} \times \frac{10}{7} = \frac{210}{175} = \frac{6}{5} = 1\frac{1}{5}$

9. $\frac{4}{5} \div \frac{1}{6} = \frac{4}{5} \times \frac{6}{1} = \frac{24}{5} = 4\frac{4}{5}$

10. $\frac{3}{4} \div \frac{1}{3} = \frac{3}{4} \times \frac{3}{1} = \frac{9}{4} = 2\frac{1}{4}$ dozen

11. $\frac{3}{4} \div \frac{1}{8} = \frac{3}{4} \times \frac{8}{1} = \frac{24}{4} = 6$ pieces

12. $\frac{5}{8} \div \frac{1}{16} = \frac{5}{8} \times \frac{16}{1} = \frac{80}{8} = 10$ pieces

13. $\frac{2}{3} \div \frac{1}{12} = \frac{2}{3} \times \frac{12}{1} = \frac{24}{3} = 8$ pieces

LESSON 18

ON YOUR OWN (PAGE 44): $9\frac{1}{3}$ miles

PAGE 45

1. $3\frac{3}{4} \div 2\frac{1}{2} = \frac{15}{4} \div \frac{5}{2} = \frac{15}{4} \times \frac{2}{5} = \frac{30}{20} = 1\frac{10}{20} = 1\frac{1}{2}$

2. $5\frac{1}{3} \div 1\frac{1}{6} = \frac{16}{3} \div \frac{7}{6} = \frac{16}{3} \times \frac{6}{7} = \frac{96}{21} = \frac{32}{7} = 4\frac{4}{7}$

3. $1\frac{9}{10} \div 3\frac{1}{5} = \frac{19}{10} \div \frac{16}{5} = \frac{19}{10} \times \frac{5}{16} = \frac{95}{160} = \frac{19}{32}$

4. $4\frac{1}{2} \div 3\frac{1}{3} = \frac{9}{2} \div \frac{10}{3} = \frac{9}{2} \times \frac{3}{10} = \frac{27}{20} = 1\frac{7}{20}$

5. $3\frac{3}{4} \div 1\frac{2}{5} = \frac{15}{4} \div \frac{7}{5} = \frac{15}{4} \times \frac{5}{7} = \frac{75}{28} = 2\frac{19}{28}$

6. $4\frac{1}{6} \div 1\frac{7}{8} = \frac{25}{6} \div \frac{15}{8} = \frac{25}{6} \times \frac{8}{15} = \frac{200}{90} = \frac{20}{9} = 2\frac{2}{9}$

7. $1\frac{5}{12} \div 1\frac{5}{6} = \frac{17}{12} \div \frac{11}{6} = \frac{17}{12} \times \frac{6}{11} = \frac{102}{132} = \frac{51}{66} = \frac{17}{22}$

8. $2\frac{2}{25} \div 1\frac{1}{10} = \frac{52}{25} \div \frac{11}{10} = \frac{52}{25} \times \frac{10}{11} = \frac{520}{275} = 1\frac{245}{275} = 1\frac{49}{55}$

9. $8\frac{1}{3} \div 2\frac{1}{5} = \frac{25}{3} \div \frac{11}{5} = \frac{25}{3} \times \frac{5}{11} = \frac{125}{33} = 3\frac{26}{33}$

10. $2\frac{3}{4} \div 1\frac{1}{2} = \frac{11}{4} \div \frac{3}{2} = \frac{11}{4} \times \frac{2}{3} = \frac{22}{12} = \frac{11}{6} = 1\frac{5}{6}$ cups

11. $12\frac{3}{8} \div 2\frac{1}{3} = \frac{99}{8} \div \frac{7}{3} = \frac{99}{8} \times \frac{3}{7} = \frac{297}{56} = 5\frac{17}{56}$ pieces

12. $6\frac{1}{4} \div 2\frac{1}{2} = \frac{25}{4} \div \frac{5}{2} = \frac{25}{4} \times \frac{2}{5} = \frac{50}{20} = \frac{5}{2} = 2\frac{1}{2}$ feet; the two full-length pieces are $2\frac{1}{2}$ feet long, the half-piece is half of $2\frac{1}{2}$ or $1\frac{1}{4}$ feet.

13. $\frac{16}{1} \div 1\frac{1}{3} = \frac{16}{1} \div \frac{4}{3} = \frac{16}{1} \times \frac{3}{4} = \frac{48}{4} = 12$ scoops

LESSON 19

ON YOUR OWN (PAGE 46): 5:7, 5 to 7, or $\frac{5}{7}$

PAGE 47

1. There are 2 juniors and 5 freshmen. So the ratio of juniors to freshmen is 2 : 5, 2 to 5, or $\frac{2}{5}$.

2. There are 2 juniors and 5 seniors. So the ratio of juniors to seniors is 2 : 5, 2 to 5, or $\frac{2}{5}$.

3. There are 6 sophomores and 5 seniors. So the ratio of sophomores to seniors is 6 : 5, 6 to 5, or $\frac{6}{5}$.

4. There are 6 sophomores and 18 riders. So the ratio of sophomores and riders is $\frac{6}{18}$. Simplified, the ratio is $\frac{6}{18} \div \frac{6}{6} = \frac{1}{3}$, 1 : 3, or 1 to 3.

5. $\frac{5}{3}$, 5 : 3, 5 to 3

6. $\frac{5}{6}$, 5 : 6, 5 to 6

7. $\frac{3}{6}$, 3 : 6, 3 to 6 or $\frac{1}{2}$, 1 : 2, 1 to 2

8. $\frac{6}{8}$, 6 : 8, 6 to 8 or $\frac{3}{4}$, 3 : 4, 3 to 4

9. The ratio of letters is 7 : 11, 7 to 11, or $\frac{7}{11}$.

10. The ratio of Kevin's laps to Sarah's laps is $\frac{25}{35}$. Simplified, the ratio is $\frac{25}{35} \div \frac{5}{5} = \frac{5}{7}$, 5 : 7, or 5 to 7.

11. The ratio of Ethan's points to Ben's points is $\frac{8}{12}$. Simplified, the ratio is $\frac{8}{12} \div \frac{4}{4} = \frac{2}{3}$, 2 : 3, or 2 to 3.

12. The ratio of winning costumes to people on the team is $\frac{7}{11}$, 7 : 11, or 7 to 11.

LESSON 20

ON YOUR OWN (PAGE 48): $\frac{35 \text{ flips}}{1 \text{ minute}}$
PAGE 49

1. $\frac{10 \text{ miles}}{6 \text{ hours}} = \frac{10 \text{ miles} \div 2}{6 \text{ hours} \div 2} = \frac{5 \text{ miles}}{3 \text{ hours}}$
The rate is 5 miles in 3 hours.

2. $\frac{20 \text{ dollars}}{6 \text{ books}} = \frac{20 \text{ dollars} \div 2}{6 \text{ books} \div 2} = \frac{10 \text{ dollars}}{3 \text{ books}}$
The rate is $10 for every 3 books.

3. $\frac{9 \text{ free throws}}{24 \text{ attempts}} = \frac{9 \text{ free throws} \div 3}{24 \text{ attempts} \div 3} = \frac{3 \text{ free throws}}{8 \text{ attempts}}$
The rate is 3 free throws for every 8 attempts.

4. $\frac{50 \text{ liters}}{4 \text{ minutes}} = \frac{50 \text{ liters} \div 2}{4 \text{ minutes} \div 2} = \frac{25 \text{ liters}}{2 \text{ minutes}}$
The rate is 25 liters for every 2 minutes.

5. $\frac{\$32}{4 \text{ hours}} = \frac{\$32 \div 4}{4 \text{ hours} \div 4} = \frac{\$8}{1 \text{ hour}}$
The rate is $8 in an hour.

6. $\frac{5 \text{ tickets}}{\$80} = \frac{5 \text{ tickets} \div 5}{\$80 \div 5} = \frac{1 \text{ ticket}}{\$16}$
The rate is 1 ticket for $16.

7. $\frac{6 \text{ cups flour}}{4 \text{ eggs}} = \frac{6 \text{ cups flour} \div 2}{4 \text{ eggs} \div 2} = \frac{3 \text{ cups flour}}{2 \text{ eggs}}$
The rate is 3 cups of flour for 2 eggs.

8. $\frac{\$300}{4 \text{ tires}} = \frac{\$300 \div 4}{4 \text{ tires} \div 4} = \frac{\$75}{1 \text{ tire}}$
The rate is $75 for each tire.

9. $\frac{10 \text{ laps}}{5 \text{ minutes}} = \frac{10 \text{ laps} \div 5}{5 \text{ minutes} \div 5} = \frac{2 \text{ laps}}{1 \text{ minute}}$
The rate is 2 laps in 1 minute.

10. $\frac{80 \text{ miles}}{2 \text{ hours}} = \frac{80 \text{ miles} \div 2}{2 \text{ hours} \div 2} = \frac{40 \text{ miles}}{1 \text{ hour}}$

11. $\frac{10 \text{ rolls}}{4 \text{ hours}} = \frac{10 \text{ rolls} \div 2}{4 \text{ hours} \div 2} = \frac{5 \text{ rolls}}{2 \text{ hours}}$

12. $\frac{15 \text{ laps}}{20 \text{ minutes}} = \frac{15 \text{ laps} \div 5}{20 \text{ minutes} \div 5} = \frac{3 \text{ laps}}{4 \text{ minutes}}$

13. $\frac{\$105}{14 \text{ days}} = \frac{\$105 \div 7}{14 \text{ days} \div 7} = \frac{\$15}{2 \text{ days}}$

14. $\frac{30 \text{ squirrels}}{4 \text{ hours}} = \frac{30 \text{ squirrels} \div 2}{4 \text{ hours} \div 2} = \frac{15 \text{ squirrels}}{2 \text{ hours}}$

15. $\frac{30 \text{ gallons}}{510 \text{ miles}} = \frac{30 \text{ gallons} \div 30}{510 \text{ miles} \div 30} = \frac{1 \text{ gallon}}{17 \text{ miles}}$

LESSON 21

ON YOUR OWN (PAGE 50): 50 miles per hour
PAGE 51

1. $\frac{60 \text{ miles}}{4 \text{ hours}} = \frac{60 \text{ miles} \div 4}{4 \text{ hours} \div 4} = \frac{15 \text{ miles}}{1 \text{ hour}}$
The rate is 15 miles per hour.

2. $\frac{\$45}{5 \text{ hours}} = \frac{\$45 \div 5}{5 \text{ hours} \div 5} = \frac{\$9}{\text{hour}}$
The rate is $9 per hour.

3. $\frac{200 \text{ words}}{4 \text{ minutes}} = \frac{200 \text{ words} \div 4}{4 \text{ minutes} \div 4} = \frac{50 \text{ words}}{\text{minute}}$
The rate is 50 words per minute.

4. $\frac{88 \text{ miles}}{2 \text{ hours}} = \frac{88 \text{ miles} \div 2}{2 \text{ hours} \div 2} = \frac{44 \text{ miles}}{\text{hour}}$
The rate is 44 miles per hour.

5. $\frac{\$1.80}{6 \text{ cans}} = \frac{\$1.80 \div 6}{6 \text{ cans} \div 6} = \frac{\$0.30}{\text{can}}$
The rate is 30 cents per can.

6. $\frac{36 \text{ grams}}{6 \text{ servings}} = \frac{36 \text{ grams} \div 6}{6 \text{ servings} \div 6} = \frac{6 \text{ grams}}{\text{serving}}$
The rate is 6 grams of fat per serving.

7. $\frac{3,000 \text{ meters}}{10 \text{ minutes}} = \frac{3,000 \text{ meters} \div 10}{10 \text{ minutes} \div 10} = \frac{300 \text{ meters}}{\text{minute}}$
The rate is 300 meters per minute.

8. $\frac{144 \text{ miles}}{6 \text{ gallons}} = \frac{144 \text{ miles} \div 6}{6 \text{ gallons} \div 6} = \frac{24 \text{ miles}}{\text{gallon}}$
The rate is 24 miles per gallon.

9. $\frac{90 \text{ pages}}{180 \text{ minutes}} = \frac{90 \text{ pages} \div 180}{180 \text{ minutes} \div 180} = \frac{0.5 \text{ page}}{\text{minute}}$
The rate is 0.5 page per minute.

10. $\frac{\$4.80}{12 \text{ bottles}} = \frac{\$4.80 \div 12}{12 \text{ bottles} \div 12} = \frac{\$0.40}{\text{bottle}}$
The rate is 40 cents per bottle.

11. $\frac{60 \text{ pages}}{30 \text{ minutes}} = \frac{60 \text{ pages} \div 30}{30 \text{ minutes} \div 30} = \frac{2 \text{ pages}}{\text{minute}}$
The rate is 2 pages per minute.

12. $\frac{500 \text{ words}}{10 \text{ minutes}} = \frac{500 \text{ words} \div 10}{10 \text{ minutes} \div 10} = \frac{50 \text{ words}}{\text{minute}}$
The rate is 50 words per minute.

13. $\frac{\$9}{6 \text{ slices}} = \frac{\$9 \div 6}{6 \text{ slices} \div 6} = \frac{\$1.50}{\text{slice}}$
The rate is $1.50 per slice.

14. $\frac{\$350}{4 \text{ rooms}} = \frac{\$350 \div 4}{4 \text{ rooms} \div 4} = \frac{\$87.50}{\text{room}}$
The rate is $87.50 per room.

15. $\frac{\$4}{5 \text{ pounds}} = \frac{\$4 \div 5}{5 \text{ pounds} \div 5} = \frac{\$0.80}{\text{pound}}$
The rate is 80 cents per pound.

LESSON 22

ON YOUR OWN (PAGE 52): No

PAGE 53

1. $2 \times 18 = 36; 3 \times 12 = 36$
   Because $36 = 36$, the ratios are a proportion.
2. $18 \times 3 = 54; 6 \times 8 = 48$
   Because $54 \neq 48$, the ratios are not proportional.
3. $2 \times 20 = 40; 8 \times 5 = 40$
   Because $40 = 40$, the ratios are a proportion.
4. The ratios are $\frac{6}{8}$ and $\frac{12}{16}$.
   $6 \times 16 = 96; 8 \times 12 = 96$
   Because $96 = 96$, the ratios are a proportion.
5. The ratios are $\frac{4}{9}$ and $\frac{12}{26}$.
   $4 \times 26 = 104; 9 \times 12 = 108$
   Because $104 \neq 108$, the ratios are not proportional.
6. $3 \times 6 = 18; 7 \times 14 = 98$
   Because $18 \neq 98$, the ratios are not proportional.
7. $7 \times 6 = 42; 2 \times 21 = 42$
   Because $42 = 42$, the ratios are a proportion.
8. The ratios are $\frac{9}{5}$ and $\frac{27}{9}$.
   $9 \times 9 = 81; 5 \times 27 = 135$
   Because $81 \neq 135$, the ratios are not proportional.
9. $2 \times 80 = 160; 8 \times 20 = 160$
   Because $160 = 160$, the ratios are proportional.
10. $40 \times 6 = 240; 4 \times 60 = 240$
    Because $240 = 240$, the ratios are proportional. They are paid at the same rate.
11. The ratios are $\frac{3}{10}$ and $\frac{5}{20}$.
    $3 \times 20 = 60; 10 \times 5 = 50$
    Because $60 \neq 50$, the ratios are not proportional. They do not save money at the same rate.
12. The ratios are $\frac{6}{15}$ and $\frac{9}{21}$.
    $6 \times 21 = 126; 9 \times 15 = 135$
    Because $126 \neq 135$, the ratios are not proportional. They do not complete passes at the same rate.
13. The ratios are $\frac{50}{1}$ and $\frac{200}{4}$.
    $50 \times 4 = 200; 1 \times 200 = 200$
    Because $200 = 200$, the ratios are proportional. Their typing speeds are the same.
14. The ratios are $\frac{0.5}{10}$ and $\frac{1}{25}$.
    $0.5 \times 25 = 12.5; 10 \times 1 = 10$
    Because $12.5 \neq 10$, the ratios are not proportional. They do not walk at the same rate.
15. The ratios are $\frac{8}{10}$ and $\frac{15}{20}$.
    $8 \times 20 = 160; 10 \times 15 = 150$
    Because $160 \neq 150$, the ratios are not proportional. They do not have the same ratio of correct answers.

LESSON 23

ON YOUR OWN (PAGE 54): 12 inches

PAGE 55

1. $4 \times n = 3 \times 12$
   $4n = 36$
   $\frac{4n}{4} = \frac{36}{4}$
   $n = 9$
   $\frac{3}{4} = \frac{9}{12}$
2. $6 \times n = 4 \times 9$
   $6n = 36$
   $\frac{6n}{6} = \frac{36}{6}$
   $n = 6$
   $\frac{4}{6} = \frac{6}{9}$
3. $16 \times n = 4 \times 8$
   $16n = 32$
   $\frac{16n}{16} = \frac{32}{16}$
   $n = 2$
   $\frac{8}{16} = \frac{2}{4}$
4. $6 \times x = 5 \times 30$
   $6x = 150$
   $\frac{6x}{6} = \frac{150}{6}$
   $x = 25$
   $\frac{25}{30} = \frac{5}{6}$
5. $30 \times n = 18 \times 10$
   $30n = 180$
   $\frac{30n}{30} = \frac{180}{30}$
   $n = 6$
   $\frac{18}{30} = \frac{6}{10}$
6. $8 \times n = 40 \times 5$
   $8n = 200$
   $\frac{8n}{8} = \frac{200}{8}$
   $n = 25$
   $\frac{40}{25} = \frac{8}{5}$
7. $5 \times x = 4 \times 25$
   $5x = 100$
   $\frac{5x}{5} = \frac{100}{5}$
   $x = 20$
   $\frac{5}{25} = \frac{4}{20}$

8. $3 \times y = 8 \times 21$

$3y = 168$

$\dfrac{3y}{3} = \dfrac{168}{3}$

$y = 56$

$\dfrac{3}{8} = \dfrac{21}{56}$

9. $4 \times y = 11 \times 20$

$4y = 220$

$\dfrac{4y}{4} = \dfrac{220}{4}$

$y = 55$

$\dfrac{55}{20} = \dfrac{11}{4}$

10. $\dfrac{20}{\$120} = \dfrac{15}{n}$

$20 \times n = \$120 \times 15$

$20n = \$1,800$

$\dfrac{20n}{20} = \dfrac{\$1,800}{20}$

$n = \$90$

11. $\dfrac{15}{10} = \dfrac{57}{n}$

$15 \times n = 10 \times 57$

$15n = 570$

$\dfrac{15n}{15} = \dfrac{570}{15}$

$n = 38$ rolls of black and white

12. $\dfrac{35}{25} = \dfrac{n}{10}$

$25 \times n = 10 \times 35$

$25n = 350$

$\dfrac{25n}{25} = \dfrac{350}{25}$

$n = 14$ laps

13. $\dfrac{140}{4} = \dfrac{n}{6}$

$4 \times n = 140 \times 6$

$4n = 840$

$\dfrac{4n}{4} = \dfrac{840}{4}$

$n = 210$ minutes

14. $\dfrac{80}{5} = \dfrac{n}{20}$

$5 \times n = 80 \times 20$

$5n = 1,600$

$\dfrac{5n}{5} = \dfrac{1,600}{5}$

$n = 320$ points

15. $\dfrac{324}{60} = \dfrac{n}{20}$

$60 \times n = 20 \times 324$

$60n = 6,480$

$\dfrac{60n}{60} = \dfrac{6,480}{60}$

$n = \$108$

ON YOUR OWN (PAGE 56): 225 words
PAGE 57

1. $\dfrac{3}{2} = \dfrac{n}{6}$

$3 \times 6 = 2 \times n$

$18 = 2n$

$\dfrac{18}{2} = \dfrac{2n}{2}$

$n = 9$ hours

2. $\dfrac{42}{3} = \dfrac{147}{n}$

$42 \times n = 3 \times 147$

$42n = 441$

$\dfrac{42n}{42} = \dfrac{441}{42}$

$n = 10.5$ hours

3. $\dfrac{20}{5} = \dfrac{n}{30}$

$5 \times n = 20 \times 30$

$5n = 600$

$\dfrac{5n}{5} = \dfrac{600}{5}$

$n = 120$ minutes

4. $\dfrac{7}{14} = \dfrac{n}{8}$

$14 \times n = 7 \times 8$

$14n = 56$

$\dfrac{14n}{14} = \dfrac{56}{14}$

$n = 4$ goals

5. $\dfrac{18}{1} = \dfrac{n}{9}$

$1 \times n = 18 \times 9$

$n = 162$ meetings

6. $\dfrac{40}{5} = \dfrac{n}{40}$

$5 \times n = 40 \times 40$

$5n = 1,600$

$\dfrac{5n}{5} = \dfrac{1,600}{5}$

$n = 320$ people

7. $\dfrac{171}{3} = \dfrac{n}{1}$

$3 \times n = 171 \times 1$

$3n = 171$

$\dfrac{3n}{3} = \dfrac{171}{3}$

$n = 57$ words

8. $\dfrac{2}{7} = \dfrac{30}{n}$

$2 \times n = 7 \times 30$

$2n = 210$

$\dfrac{2n}{2} = \dfrac{210}{2}$

$n = 105$ students

9. $\dfrac{\$390}{30 \text{ hours}} = \dfrac{n}{1 \text{ hour}}$

$30 \times n = 390 \times 1$

$30n = 390$

$\dfrac{30n}{30} = \dfrac{390}{30}$

$n = \$13$ per hour

10. $\dfrac{360 \text{ calories}}{6 \text{ servings}} = \dfrac{n}{1 \text{ serving}}$

$6 \times n = 1 \times 360$

$6n = 360$

$\dfrac{6n}{6} = \dfrac{360}{6}$

$n = 60$ calories per serving

11. $\dfrac{8 \text{ laps}}{20 \text{ minutes}} = \dfrac{n}{60 \text{ minutes}}$

$20 \times n = 8 \times 60$

$20n = 480$

$\dfrac{20n}{20} = \dfrac{480}{20}$

$n = 24$ laps

12. $\dfrac{8 \text{ calls}}{12 \text{ minutes}} = \dfrac{n}{30 \text{ minutes}}$

$12 \times n = 8 \times 30$

$12n = 240$

$\dfrac{12n}{12} = \dfrac{240}{12}$

$n = 20 \text{ calls}$

13. $\dfrac{36 \text{ problems}}{60 \text{ minutes}} = \dfrac{n}{30 \text{ minutes}}$

$60 \times n = 36 \times 30$

$60n = 1{,}080$

$\dfrac{60n}{60} = \dfrac{1{,}080}{60}$

$n = 18 \text{ problems}$

14. $\dfrac{12 \text{ servings}}{2 \text{ cups carrots}} = \dfrac{n}{0.5 \text{ cup carrots}}$

$2 \times n = 12 \times 0.5$

$2n = 6$

$\dfrac{2n}{2} = \dfrac{6}{2}$

$n = 3 \text{ servings}$

LESSON 25

ON YOUR OWN (PAGE 58): $12.50
PAGE 59

1. $\dfrac{3 \text{ onions}}{1 \text{ pound}} = \dfrac{12 \text{ onions}}{n}$

$3 \times n = 12 \times 1$

$3n = 12$

$\dfrac{3n}{3} = \dfrac{12}{3}$

$n = 4$

Twelve onions weigh 4 pounds.

2. $\dfrac{\$6}{5 \text{ fish}} = \dfrac{n}{10 \text{ fish}}$

$5 \times n = 10 \times 6$

$5n = 60$

$\dfrac{5n}{5} = \dfrac{60}{5}$

$n = 12$

Ten fish cost $12.

3. $\dfrac{\$4.80}{12 \text{ pens}} = \dfrac{n}{10 \text{ pens}}$

$12 \times n = 4.80 \times 10$

$12n = 48.0$

$\dfrac{12n}{12} = \dfrac{48}{12}$

$n = 4$

Ten pens cost $4.00.

4. $\dfrac{\$10}{6 \text{ roses}} = \dfrac{\$15}{n}$

$10 \times n = 6 \times 15$

$10n = 90$

$\dfrac{10n}{10} = \dfrac{90}{10}$

$n = 9$

Nine roses cost $15.

5. $\dfrac{35 \text{ miles}}{2 \text{ hours}} = \dfrac{n}{6 \text{ hours}}$

$2 \times n = 6 \times 35$

$2n = 210$

$\dfrac{2n}{2} = \dfrac{210}{2}$

$n = 105$

You can bike 105 miles in 6 hours.

6. $\dfrac{8 \text{ people}}{0.75 \text{ cup}} = \dfrac{12 \text{ people}}{n}$

$8 \times n = 0.75 \times 12$

$8n = 9$

$\dfrac{8n}{8} = \dfrac{9}{8}$

$n = \dfrac{9}{8} = 1\dfrac{1}{8}$

You will need $1\dfrac{1}{8}$ cups water.

7. $\dfrac{24 \text{ calls}}{2 \text{ hours}} = \dfrac{n}{0.5 \text{ hours}}$

$2 \times n = 24 \times 0.5$

$2n = 12$

$\dfrac{2n}{2} = \dfrac{12}{2}$

$n = 6$

You can make 6 calls.

8. $\dfrac{12 \text{ bananas}}{3 \text{ pounds}} = \dfrac{2 \text{ bananas}}{n}$

$12 \times n = 3 \times 2$

$12n = 6$

$\dfrac{12n}{12} = \dfrac{6}{12}$

$n = 0.5$

Two bananas weigh half a pound.

9. $\dfrac{10 \text{ songs}}{20 \text{ minutes}} = \dfrac{n}{5 \text{ minutes}}$

$20 \times n = 10 \times 5$

$20n = 50$

$\dfrac{20n}{20} = \dfrac{50}{20}$

$n = 2.5$

They play 2.5 songs.

10. $\dfrac{4 \text{ black}}{20 \text{ marbles}} = \dfrac{n}{200 \text{ marbles}}$

$20 \times n = 4 \times 200$

$20n = 800$

$\dfrac{20n}{20} = \dfrac{800}{20}$

$n = 40$

Forty of the marbles are black.

LESSON 26

ON YOUR OWN (PAGE 60): 22%
PAGE 61

1. $\frac{44}{100} = 44\%$ 2. $\frac{58}{100} = 58\%$
3. $\frac{91}{100} = 91\%$ 4. $\frac{30}{100} = 30\%$
5. $\frac{75}{100} = 75\%$ 6. $\frac{85}{100} = 85\%$
7. $\frac{66}{100} = 66\%$ 8. $\frac{92}{100} = 92\%$
9. $\frac{53}{100} = 53\%$ 10. $\frac{14}{100} = 14\%$
11. $\frac{50}{100} = 50\%$ 12. $\frac{17}{100} = 17\%$
13. $\frac{23}{100} = 23\%$ 14. $\frac{46}{100} = 46\%$

LESSON 27

ON YOUR OWN (PAGE 62): 60%
PAGE 63

1. $\frac{4}{5} = 4 \div 5 = 0.80 = 80\%$
2. $\frac{9}{10} = 9 \div 10 = 0.90 = 90\%$
3. $\frac{11}{20} = 11 \div 20 = 0.55 = 55\%$
4. $\frac{3}{50} = 3 \div 50 = 0.06 = 6\%$
5. $\frac{3}{8} = 3 \div 8 = 0.375 = 37.5\%$
6. $\frac{22}{100} = 22 \div 100 = 22\%$
7. $\frac{64}{200} = 64 \div 200 = 0.32 = 32\%$
8. $\frac{14}{20} = 14 \div 20 = 0.70 = 70\%$
9. $\frac{80}{400} = 80 \div 400 = 0.20 = 20\%$
10. $\frac{17}{20} = 17 \div 20 = 0.85 = 85\%$
11. $\frac{40}{50} = 40 \div 50 = 0.80 = 80\%$
12. $\frac{28}{40} = 28 \div 40 = 0.70 = 70\%$
13. $\frac{7}{8} = 7 \div 8 = 0.875 = 87.5\%$
14. $\frac{24}{400} = 24 \div 400 = 0.06 = 6\%$
15. $\frac{16}{80} = 16 \div 80 = 0.20 = 20\%$

LESSON 28

ON YOUR OWN (PAGE 64): $\frac{17}{20}$
PAGE 65

1. $55\% = \frac{55}{100} = \frac{55 \div 5}{100 \div 5} = \frac{11}{20}$
2. $95\% = \frac{95}{100} = \frac{95 \div 5}{100 \div 5} = \frac{19}{20}$
3. $15\% = \frac{15}{100} = \frac{15 \div 5}{100 \div 5} = \frac{3}{20}$
4. $44\% = \frac{44}{100} = \frac{44 \div 4}{100 \div 4} = \frac{11}{25}$
5. $83\% = \frac{83}{100}$
6. $28\% = \frac{28}{100} = \frac{28 \div 4}{100 \div 4} = \frac{7}{25}$
7. $8\% = \frac{8}{100} = \frac{8 \div 4}{100 \div 4} = \frac{2}{25}$

8. $34\% = \frac{34}{100} = \frac{34 \div 2}{100 \div 2} = \frac{17}{50}$
9. $68\% = \frac{68}{100} = \frac{68 \div 4}{100 \div 4} = \frac{17}{25}$
10. $45\% = \frac{45}{100} = \frac{45 \div 5}{100 \div 5} = \frac{9}{20}$
11. $43\% = \frac{43}{100}$
12. $15\% = \frac{15}{100} = \frac{15 \div 5}{100 \div 5} = \frac{3}{20}$
13. $68\% = \frac{68}{100} = \frac{68 \div 4}{100 \div 4} = \frac{17}{25}$
14. $29\% = \frac{29}{100}$
15. $28\% = \frac{28}{100} = \frac{28 \div 4}{100 \div 4} = \frac{7}{25}$

LESSON 29

ON YOUR OWN (PAGE 66): 80%
PAGE 67

1. $0.06 = 6\%$ 2. $0.64 = 64\%$
3. $0.052 = 5.2\%$ 4. $0.888 = 88.8\%$
5. $0.7 = 70\%$ 6. $0.005 = 0.5\%$
7. $0.0024 = 0.24\%$ 8. $0.908 = 90.8\%$
9. $1.4 = 140\%$ 10. $1.063 = 106.3\%$
11. $0.45 = 45\%$ 12. $0.95 = 95\%$
13. $0.35 = 35\%$ 14. $0.05 = 5\%$
15. $0.652 = 65.2\%$
   $100.0\% - 65.2\% = 34.8\%$
16. $0.98 = 98\%$

LESSON 30

ON YOUR OWN (PAGE 68): 0.255
PAGE 69

1. $85\% = 0.85$ 2. $65\% = 0.65$
3. $12\% = 0.12$ 4. $48\% = 0.48$
5. $73\% = 0.73$ 6. $29\% = 0.29$
7. $4\% = 0.04$ 8. $34.7\% = 0.347$
9. $150\% = 1.50$ 10. $2.5\% = 0.025$
11. $3.1\% = 0.031$ 12. $0.3\% = 0.003$
13. $74\% = 0.74$ teens   14. $40\% = 0.40$ moviegoers
15. $48\% = 0.48$ teens   16. $5\% = 0.05$ teens
17. $100.0\% - 62.5\% = 37.5\%$
   $37.5\% = 0.375$ movies
18. $120\% = 1.20$ effort; Accept any reasonable answer.

LESSON 31

ON YOUR OWN (PAGE 70): 39 students
PAGE 71

1. $82\% \times 50 = .82 \times 50 = 41$
2. $31\% \times 66 = 0.31 \times 66 = 20.46$
3. $10\% \times 45 = 0.10 \times 45 = 4.5$
4. $50\% \times 600 = 0.50 \times 600 = 300$
5. $23\% \times 90 = 0.23 \times 90 = 20.7$
6. $2.5\% \times 22 = 0.025 \times 22 = 0.55$
7. $60\% \times 60 = 0.60 \times 60 = 36$
8. $64.7\% \times 12 = 0.647 \times 12 = 7.764$
9. $150\% \times 4 = 1.50 \times 4 = 6$
10. $28\% \times 150 = 0.28 \times 150 = 42$ seniors
11. $75\% \times 32 = 0.75 \times 32 = 24$ actors
12. $62.5\% \times 40 = 0.625 \times 40 = 25$ stores
13. $37.5\% \times 24 = 0.375 \times 24 = 9$ fly balls
14. $70\% \times 140 = 0.70 \times 140 = 98$ houses
15. $80\% \times \$250,000 = 0.80 \times \$250,000 = \$200,000$

LESSON 32

ON YOUR OWN (PAGE 72): 50
PAGE 73

1. $42 \div 50\% = 42 \div 0.50 = 84$
2. $48 \div 40\% = 48 \div 0.40 = 120$
3. $90 \div 30\% = 90 \div 0.30 = 300$
4. $15 \div 125\% = 15 \div 1.25 = 12$
5. $25\% \times n = 20$
   $n = 20 \div 25\% = 20 \div 0.25 = 80$
6. $120\% \times n = 18$
   $n = 18 \div 120\% = 18 \div 1.20 = 15$
7. $125\% \times n = 120$
   $n = 120 \div 125\% = 120 \div 1.25 = 96$
8. $25\% \times n = 72$
   $n = 72 \div 25\% = 72 \div 0.25 = 288$
9. $24 \div 30\% = 24 \div 0.30 = 80$ students applied
10. $32 \div 16\% = 32 \div 0.16 = 200$ businesses
11. $300 \div 6\% = 300 \div 0.06 = 5,000$ applied
12. $330 \div 165\% = 330 \div 1.65 = 200$ deer
13. $45 \div 37.5\% = 45 \div 0.375 = 120$ parents
14. $75 \div 12.5\% = 75 \div 0.125 = 600$ students

LESSON 33

ON YOUR OWN (PAGE 74): 25%
PAGE 75

1. $(60 \div 240) \times 100\% = 0.25 \times 100\% = 25\%$
2. $(90 \div 240) \times 100\% = 0.375 \times 100\% = 37.5\%$
3. $(72 \div 288) \times 100\% = 0.25 \times 100\% = 25\%$
4. $(140 \div 200) \times 100\% = 0.70 \times 100\% = 70\%$
5. $(18 \div 360) \times 100\% = 0.05 \times 100\% = 5\%$
6. $(13 \div 80) \times 100\% = 0.1625 \times 100\% = 16.25\%$
7. $(27 \div 50) \times 100\% = 0.54 \times 100\% = 54\%$
8. $(6 \div 40) \times 100\% = 0.15 \times 100\% = 15\%$
9. $(40 \div 25) \times 100\% = 1.6 \times 100\% = 160\%$
10. $(12 \div 75) \times 100\% = 0.16 \times 100\% = 16\%$
11. $(400 \div 1,200) \times 100\% = 0.333 \times 100\% = 33.3\%$
12. $(36 \div 48) \times 100\% = 0.75 \times 100\% = 75\%$
13. $(144 \div 32) \times 100\% = 4.5 \times 100\% = 450\%$
14. $(56 \div 224) \times 100\% = 0.25 \times 100\% = 25\%$

LESSON 34

ON YOUR OWN (PAGE 76): 25%
PAGE 77

1. $n \times 60 = 24$
2. $n = 0.32 \times 80$
3. $n \times 64 = 12$
4. $0.80 \times n = 16$
5. $n \times 240 = 180$
6. $n = 0.28 \times 54$
7. $0.62 \times n = 16$
8. $n \times 20 = 60$
9. $n = 0.02 \times 10$
10. $0.10 \times n = 12$
11. $n \times 3.6 = 1.8$
12. $n = 0.022 \times 200$
13. $0.44 \times n = 22$
14. $n \times 68 = 16$
15. $n = 0.0232 \times 50$
16. $0.75 \times n = 33$

LESSON 35

ON YOUR OWN (PAGE 78): 1,800 marchers
PAGE 79

1. 44 is 20% of what number?
   $44 = 0.2 \times n; n = 220$ houses
2. 30% of 1,340 is what number?
   $0.30 \times 1,340 = n; n = \$402$
3. What percent of 50 (28 + 22) is 22?
   $n \times 50 = 22; n = 44\%$

4. 55 is 11% of what number?

$0.11 \times n = 55; n = 500$ postcards

5. 85% of 8 is what number?

$0.85 \times 8 = n; n = 6.8$ tons

6. What percent of 50 is 125 $(175 - 50)$?

$n \times 50 = 125; n = 250\%$

7. 0.2% of what number is 18?

$0.002 \times n = 18; n = 9,000$ students

8. What number is 28% of 150?

$n = 0.28 \times 150; n = 42$ players

9. What percent of 623 is 386?

$n \times 623 = 386; n = 62\%$

10. 94.5% of what number is 821?

$0.945 \times n = 821; n = 868.8 = 869$ people

LESSON 36

ON YOUR OWN (PAGE 80): 12.5%

PAGE 81

1. $60 - 57 = 3$ fewer teachers

$3 = n \times 60$

$\dfrac{3}{60} = \dfrac{60n}{60}$

$n = \dfrac{3}{60} = \dfrac{1}{20} = 0.05 = 5\%$

The percent decrease is 5%.

2. $\$60 - \$45 = \$15$ decrease

$15 = n \times 60$

$\dfrac{15}{60} = \dfrac{60n}{60}$

$n = \dfrac{15}{60} = \dfrac{1}{4} = 0.25 = 25\%$

The percent decrease is 25%.

3. $850 - 800 = 50$ seats increase

$50 = n \times 800$

$\dfrac{50}{800} = \dfrac{800n}{800}$

$n = \dfrac{50}{800} = \dfrac{1}{16} = 0.0625 = 6.25\%$

The percent increase is 6.25%.

4. $32 - 16 = 16$ acres decrease

$16 = n \times 32$

$\dfrac{16}{32} = \dfrac{32n}{32}$

$n = \dfrac{16}{32} = \dfrac{1}{2} = 0.50 = 50\%$

The percent decrease is 50%.

5. $48 - 40 = 8$ larger

$8 = n \times 40$

$\dfrac{8}{40} = \dfrac{40n}{40}$

$n = \dfrac{8}{40} = \dfrac{1}{5} = 0.20 = 20\%$

The percent increase is 20%.

6. $50 - 42 = 8$ fewer points

$8 = n \times 50$

$\dfrac{8}{50} = \dfrac{50n}{50}$

$n = \dfrac{8}{50} = \dfrac{4}{25} = 0.16 = 16\%$

The percent decrease is 16%.

7. $\$120 - \$100 = \$20$

$20 = n \times 120$

$\dfrac{20}{120} = \dfrac{120n}{120}$

$n = \dfrac{20}{120} = \dfrac{1}{6} = 0.167 = 16.7\%$

The percent decrease is 16.7%.

8. $\$85.00 - \$76.50 = \$8.50$

$8.50 = n \times 85.00$

$\dfrac{8.50}{85.00} = \dfrac{85.00n}{85.00}$

$n = \dfrac{8.50}{85.00} = \dfrac{1}{10} = 0.10 = 10\%$

The percent decrease is 10%.

9. $50 - 45 = 5$ hours less

$5 = n \times 50$

$\dfrac{5}{50} = \dfrac{50n}{50}$

$n = \dfrac{5}{50} = \dfrac{1}{10} = 0.10 = 10\%$

The percent decrease is 10%.

10. $40 - 32 = 8$ more workers

$8 = n \times 32$

$\dfrac{8}{32} = \dfrac{32n}{32}$

$n = \dfrac{8}{32} = \dfrac{1}{4} = 0.25 = 25\%$

The percent increase is 25%.

11. $\$550 - \$500 = \$50$

$50 = n \times 500$

$\dfrac{50}{500} = \dfrac{500n}{500}$

$n = \dfrac{50}{500} = \dfrac{1}{10} = 0.10 = 10\%$

The percent increase is 10%.

12. $300 - 220 = 80$ fewer people

$80 = n \times 300$

$\dfrac{80}{300} = \dfrac{300n}{300}$

$n = \dfrac{80}{300} = \dfrac{4}{15} = 0.267 = 26.7\%$

The percent decrease is 26.7%.

13. $\$440 - \$396 = \$44$

$44 = n \times 440$

$\dfrac{44}{440} = \dfrac{440n}{440}$

$n = \dfrac{44}{440} = \dfrac{1}{10} = 0.10 = 10\%$

The percent decrease is 10%.

14. $0.240 - 0.192 = 0.048$

$0.048 = n \times 0.240$

$\dfrac{0.048}{0.240} = \dfrac{0.240n}{0.240}$

$n = \dfrac{0.048}{0.240} = \dfrac{1}{5} = 0.20 = 20\%$

The percent decrease is 20%.

LESSON 37

PAGE 83

1. $\$285 - \$240 = \$45$

   $45 = n \times 285$

   $\dfrac{45}{285} = \dfrac{285n}{285}$

   $n = \dfrac{3}{19} = 0.158 = 15.8\%$

   The percent decrease is 15.8%.

2. $\$600 - \$480 = \$120$

   $120 = n \times 600$

   $\dfrac{120}{600} = \dfrac{600n}{600}$

   $n = \dfrac{120}{600} = \dfrac{1}{5} = 0.20 = 20\%$

   The percent decrease is 20%.

3. $\$84.00 - \$75.60 = \$8.40$

   $8.40 = n \times 84.00$

   $\dfrac{8.40}{84.00} = \dfrac{84.00n}{84.00}$

   $n = \dfrac{8.40}{84.00} = \dfrac{1}{10} = 0.10 = 10\%$

   The percent decrease is 10%.

4. $\$75 - \$60 = \$15$

   $15 = n \times 60$

   $\dfrac{15}{60} = \dfrac{60n}{60}$

   $n = \dfrac{15}{60} = \dfrac{1}{4} = 0.25 = 25\%$

   The percent increase is 25%.

5. $\$26 - \$20 = \$6$

   $6 = n \times 20$

   $\dfrac{6}{20} = \dfrac{20n}{20}$

   $n = \dfrac{6}{20} = \dfrac{3}{10} = 0.30 = 30\%$

   The percent increase is 30%.

6. $\$300 - \$210 = \$90$

   $90 = n \times 300$

   $\dfrac{90}{300} = \dfrac{300n}{300}$

   $n = \dfrac{90}{300} = \dfrac{3}{10} = 0.30 = 30\%$

   The percent decrease is 30%.

7. $\$45 - \$25 = \$20$

   $20 = n \times 25$

   $\dfrac{20}{25} = \dfrac{25n}{25}$

   $n = \dfrac{20}{25} = \dfrac{4}{5} = 0.80 = 80\%$

   The percent increase is 80%.

8. $\$28 - \$24 = \$4$

   $4 = n \times 24$

   $\dfrac{4}{24} = \dfrac{24n}{24}$

   $n = \dfrac{4}{24} = \dfrac{1}{6} = 0.167 = 16.7\%$

   The percent increase is 16.7%.

9. $\$75 - \$60 = \$15$

   $15 = n \times 75$

   $\dfrac{15}{75} = \dfrac{75n}{75}$

   $n = \dfrac{15}{75} = \dfrac{1}{5} = 0.20 = 20\%$

   The percent discount is 20%.

10. $\$80 - \$65 = \$15$

    $15 = n \times 65$

    $\dfrac{15}{65} = \dfrac{65n}{65}$

    $n = \dfrac{15}{65} = \dfrac{3}{13} = 0.23 = 23\%$

    The percent markup is 20%.

11. $\$160 - \$128 = \$32$

    $32 = n \times 160$

    $\dfrac{32}{160} = \dfrac{160n}{160}$

    $n = \dfrac{32}{160} = \dfrac{1}{5} = 0.20 = 20\%$

    The percent discount is 20%.

12. $120 = n \times 300$

    $\dfrac{120}{300} = \dfrac{300n}{300}$

    $n = \dfrac{120}{300} = \dfrac{2}{5} = 0.40 = 40\%$

    The percent markup is 40%.

13. $\$480 - \$410 = \$70$

    $70 = n \times 480$

    $\dfrac{70}{480} = \dfrac{480n}{480}$

    $n = \dfrac{70}{480} = \dfrac{7}{48} = 0.146 = 14.6\%$

    The percent discount is 14.6%.

14. $\$360 - \$280 = \$80$

    $80 = n \times 360$

    $\dfrac{80}{360} = \dfrac{360n}{360}$

    $\dfrac{80}{360} = \dfrac{2}{9} = 0.222 = 22.2\%$

    The percent discount is 22.2%.

LESSON 38

PAGE 85

1. $I = \$800 \times 3\% \times 2 = \$800 \times 0.03 \times 2 = \$48$

2. $I = \$4,000 \times 4\% \times 1 = \$4,000 \times 0.04 \times 1 = \$160$

3. $I = \$500 \times 5\% \times 2 = \$500 \times 0.05 \times 2 = \$50$

4. $I = \$700 \times 3.5\% \times 0.5 = \$700 \times 0.035 \times 0.5 = \$12.25$

5. $I = \$1,000 \times 8\% \times 2 = \$1,000 \times 0.08 \times 2 = \$160$

   $T = \$1,000 + \$160 = \$1,160$

6. $I = \$900 \times 7.5\% \times 5 = \$900 \times 0.075 \times 5 = \$337.50$

   $T = \$900 + \$337.50 = \$1,237.50$

7. $I = \$1,300 \times 5.4\% \times 1 = \$1,300 \times 0.054 \times 1 = \$70.20$
$T = \$1,300 + \$70.20 = \$1,370.20$

8. $I = \$800 \times 9\% \times \frac{9}{12} = \$800 \times 0.09 \times 0.75 = \$54$
$T = \$800 + \$54 = \$854$

9. $I = \$320 \times 0.06 \times 3 = \$57.60$
$T = \$320 + \$57.60 = \$377.60$

10. $I = \$900 \times 0.05 \times 0.5 = \$22.50$

11. $I = \$1,100 \times 0.09 \times 0.75 = \$74.25$
$T = \$1,100 + \$74.25 = \$1,174.25$

12. $I = \$240 \times 0.06 \times 1.5 = \$21.60$

13. $I = \$450 \times 0.005 \times 0.75 = \$1.69$

14. $I = \$1,000 \times 0.0675 \times 2.5 = \$168.75$
$T = \$1,000 + \$168.75 = \$1,168.75$

LESSON 39

ON YOUR OWN (PAGE 86): $486.20

PAGE 87

1. $B = 600(1 + 0.04)^2 = 600(1.04)^2 = 600(1.0816) = \$648.96$

2. $B = 900(1 + 0.06)^3 = 900(1.06)^3 = 900(1.1910) = \$1,071.90$

3. $B = 900(1 + 0.04)^5 = 900(1.04)^5 = 900(1.2167) = \$1,095.03$

4. $B = 1,800(1 + 0.06)^4 = 1,800(1.06)^4 = 1,800(1.2625) = \$2,272.50$

5. $B = 2,000(1 + 0.045)^1 = 2,000(1.045) = \$2,090$

6. $B = 3,000(1 + 0.0425)^{12} = 3,000(1.0425)^{12} = 3,000(1.6478) = \$4,943.40$

7. $B = 5,000(1 + 0.04)^5 = 5,000(1.04)^5 = 5,000(1.2167) = \$6,083.50$

8. $B = 4,250(1 + 0.06)^4 = 4,250(1.06)^4 = 4,250(1.2625) = \$5,365.63$

9. $B = 500(1 + 0.08)^{15} = 500(1.08)^{15} = 500(3.1722) = \$1,586.10$

10. $B = 300(1 + 0.01)^{12} = 300(1.01)^{12} = 300(1.1268) = \$338.04$

11. $B = 1,000(1 + 0.10)^{1.5} = 1,000(1.10)^{1.5} = 1,000(1.1537) = \$1,153.70$

12. $B = 9,000(1 + 0.03)^4 = 9,000(1.03)^4 = 9,000(1.1255) = \$10,129.50$

13. $B = 1,000(1 + 0.05)^{100} = 1,000(1.05)^{100} = 1,000(131.5012) = \$131,501.20$

14. $B = 150(1 + 0.005)^{12} = 150(1.005)^{12} = 150(1.0617) = \$159.26$

Types of Fractions

Proper Fraction

A fraction in which the numerator (the top number) is less than the denominator (the bottom number).

Examples:

$\dfrac{1}{2}$, $\dfrac{2}{3}$, $\dfrac{8}{17}$, $\dfrac{4}{25}$

Improper Fraction

A fraction in which the numerator is greater than the denominator.

Examples:

$\dfrac{3}{2}$, $\dfrac{5}{3}$, $\dfrac{38}{17}$, $\dfrac{44}{25}$

Mixed Number

A number written as the sum of an integer and a proper fraction.

Examples:

$1\dfrac{1}{2}$, $3\dfrac{2}{3}$, $4\dfrac{1}{25}$, $6\dfrac{7}{10}$

Equivalent Fractions

Fractions that name the same amount.

Example:

$\dfrac{1}{2} = \dfrac{3}{6} = \dfrac{5}{10} = \dfrac{10}{20} = \dfrac{25}{50}$

Fractions with Like Denominators

Fractions in which the denominators are the same number.

Example:

$\dfrac{1}{5}$, $\dfrac{2}{5}$, $3\dfrac{1}{5}$, $\dfrac{4}{5}$, $\dfrac{7}{5}$

Fractions with Unlike Denominators

Fractions in which the denominators are not the same number.

Example:

$\dfrac{1}{5}$, $\dfrac{2}{7}$, $3\dfrac{1}{8}$, $\dfrac{4}{9}$, $\dfrac{7}{2}$

Blackline Masters
Fractions, Ratios, and Percents, SV 0436-0

Benchmark Fractions

Some fractions are used *a lot* and are helpful in picturing other fractions. These fractions are known as **benchmark fractions.**

Knowing benchmark fractions can help you:

- compare and order fractions and mixed numbers;

$$\frac{3}{4} > \frac{1}{2}$$

- round fractions and mixed numbers;

$\frac{1}{4}$ rounds down to 0 because $\frac{1}{4}$ is closer to 0 than 1.

The fractions on the number line below are benchmark fractions. They can help you estimate fractions.

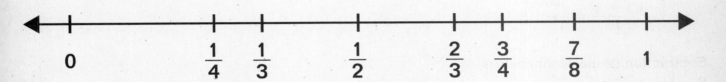

Fraction Strips

1											
$\frac{1}{2}$						$\frac{1}{2}$					
$\frac{1}{3}$				$\frac{1}{3}$				$\frac{1}{3}$			
$\frac{1}{4}$			$\frac{1}{4}$			$\frac{1}{4}$			$\frac{1}{4}$		
$\frac{1}{5}$		$\frac{1}{5}$		$\frac{1}{5}$		$\frac{1}{5}$		$\frac{1}{5}$			
$\frac{1}{6}$		$\frac{1}{6}$		$\frac{1}{6}$		$\frac{1}{6}$		$\frac{1}{6}$		$\frac{1}{6}$	
$\frac{1}{8}$	$\frac{1}{8}$	$\frac{1}{8}$	$\frac{1}{8}$	$\frac{1}{8}$	$\frac{1}{8}$	$\frac{1}{8}$	$\frac{1}{8}$				
$\frac{1}{9}$	$\frac{1}{9}$	$\frac{1}{9}$	$\frac{1}{9}$	$\frac{1}{9}$	$\frac{1}{9}$	$\frac{1}{9}$	$\frac{1}{9}$	$\frac{1}{9}$			
$\frac{1}{10}$	$\frac{1}{10}$	$\frac{1}{10}$	$\frac{1}{10}$	$\frac{1}{10}$	$\frac{1}{10}$	$\frac{1}{10}$	$\frac{1}{10}$	$\frac{1}{10}$	$\frac{1}{10}$		
$\frac{1}{12}$	$\frac{1}{12}$	$\frac{1}{12}$	$\frac{1}{12}$	$\frac{1}{12}$	$\frac{1}{12}$	$\frac{1}{12}$	$\frac{1}{12}$	$\frac{1}{12}$	$\frac{1}{12}$	$\frac{1}{12}$	$\frac{1}{12}$

Rules for Fractions

Addition (same denominators)

$$\frac{A}{B} + \frac{C}{B} = \frac{A + C}{B}$$

$$\frac{1}{8} + \frac{2}{8} = \frac{1 + 2}{8} = \frac{3}{8}$$

Addition (different denominators)

$$\frac{A}{B} + \frac{C}{D} = \frac{AD}{BD} + \frac{BC}{BD} = \frac{AD + BC}{BD}$$

$$\frac{1}{4} + \frac{1}{3} = \frac{1 \times 3}{4 \times 3} + \frac{4 \times 1}{4 \times 3}$$

$$= \frac{3}{12} + \frac{4}{12}$$

$$= \frac{3 + 4}{12} = \frac{7}{12}$$

Subtraction (same denominators)

$$\frac{A}{B} - \frac{C}{B} = \frac{A - C}{B}$$

$$\frac{3}{4} - \frac{1}{4} = \frac{3 - 1}{4} = \frac{2}{4} = \frac{1}{2}$$

Subtraction (different denominators)

$$\frac{A}{B} - \frac{C}{D} = \frac{AD}{BD} - \frac{BC}{BD} = \frac{AD - BC}{BD}$$

$$\frac{3}{4} - \frac{1}{3} = \frac{3 \times 3}{4 \times 3} - \frac{4 \times 1}{4 \times 3}$$

$$= \frac{9}{12} - \frac{4}{12}$$

$$= \frac{9 - 4}{12} = \frac{5}{12}$$

Multiplication

$$\frac{A}{B} \times \frac{C}{D} = \frac{AC}{BD}$$

$$\frac{1}{2} \times \frac{3}{4} = \frac{1 \times 3}{2 \times 4} = \frac{3}{8}$$

Division

$$\frac{A}{B} \div \frac{C}{D} = \frac{A}{B} \times \frac{D}{C} = \frac{AD}{BC}$$

$$\frac{7}{8} \div \frac{1}{3} = \frac{7}{8} \times \frac{3}{1} = \frac{7 \times 3}{8 \times 1} = \frac{21}{8} = 2\frac{5}{8}$$

Estimation

Estimation is an important math tool that enables you to answer questions such as the following:

• What is the approximate answer to this problem?

• Am I using the right operative to solve this problem?

Using Estimation with Fractions and Decimals

The following examples show you how you can round each fraction or decimal to a whole number before beginning a calculation.

Add	Estimate
$5\frac{2}{3}$ $+2\frac{1}{4}$	6 $+2$ 8
Subtract	**Estimate**
$8\frac{3}{4}$ $-3\frac{1}{8}$	9 -3 6
Multiply	**Estimate**
$4\frac{7}{8} \times 3\frac{1}{5}$	$5 \times 3 = 15$
Divide	**Estimate**
$9\frac{1}{3} \div 2\frac{3}{4}$	$9 \div 3 = 3$

Equivalent Fractions, Decimals, and Percents

Fraction	Decimal	Percent
$\frac{1}{2}$	0.5	50%
$\frac{2}{2} = 1$	1.0	100%

Fraction	Decimal	Percent
$\frac{1}{3}$	$0.33\overline{3}$	$33.\overline{3}\%$
$\frac{2}{3}$	$0.66\overline{6}$	$66.\overline{6}\%$
$\frac{3}{3} = 1$	1.0	100%

Fraction	Decimal	Percent
$\frac{1}{4}$	0.25	25%
$\frac{2}{4} = \frac{1}{2}$	0.5	50%
$\frac{3}{4}$	0.75	75%
$\frac{4}{4} = 1$	1.0	100%

Fraction	Decimal	Percent
$\frac{1}{5}$	0.2	20%
$\frac{2}{5}$	0.4	40%
$\frac{3}{5}$	0.6	60%
$\frac{4}{5}$	0.8	80%
$\frac{5}{5} = 1$	1.0	100%

Fraction	Decimal	Percent
$\frac{1}{6}$	$0.16\overline{6}$	$16.\overline{6}\%$
$\frac{2}{6} = \frac{1}{3}$	$0.33\overline{3}$	$33.\overline{3}\%$
$\frac{3}{6} = \frac{1}{2}$	0.5	50%
$\frac{4}{6} = \frac{2}{3}$	$0.66\overline{6}$	$66.\overline{6}\%$
$\frac{5}{6}$	$0.83\overline{3}$	$83.\overline{3}\%$
$\frac{6}{6} = 1$	1.0	100%

Fraction	Decimal	Percent
$\frac{1}{8}$	0.125	12.5%
$\frac{2}{8} = \frac{1}{4}$	0.25	25%
$\frac{3}{8}$	0.375	37.5%
$\frac{4}{8} = \frac{1}{2}$	0.5	50%
$\frac{5}{8}$	0.625	62.5%
$\frac{6}{8} = \frac{3}{4}$	0.75	75%
$\frac{7}{8}$	0.875	87.5%
$\frac{8}{8} = 1$	1.0	100%

Fraction	Decimal	Percent
$\frac{1}{10}$	0.1	10%
$\frac{2}{10} = \frac{1}{5}$	0.2	20%
$\frac{3}{10}$	0.3	30%
$\frac{4}{10} = \frac{2}{5}$	0.4	40%
$\frac{5}{10} = \frac{1}{2}$	0.5	50%
$\frac{6}{10} = \frac{3}{5}$	0.6	60%
$\frac{7}{10}$	0.7	70%
$\frac{8}{10} = \frac{4}{5}$	0.8	80%
$\frac{9}{10}$	0.9	90%
$\frac{10}{10} = 1$	1.0	100%

Fraction	Decimal	Percent
$\frac{1}{100}$	0.01	1%
1	1.0	100%

Percents and Problem Solving

Find the Part

Given: the percent and the whole

Equation: Whole $\times$ Percent = Part

Example: Find 30% of 120.

$120 \times 0.30 = 36$

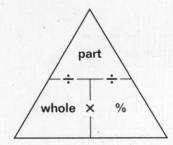

Find the Percent

Given: the part and the whole

Equation: (Part $\div$ Whole) $\times$ 100 = Percent

Example: 50 is what percent of 250?

$(50 \div 250) \times 100 = 20\%$

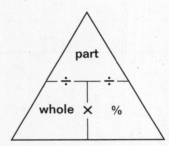

Find the Whole

Given: the percent and the part

Equation: Part $\div$ Percent = Whole

Example: 36 is 75% of what number?

$36 \div 0.75 = 48$

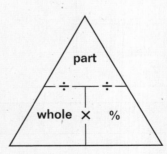

Simple and Compound Interest

Simple Interest

Simple Interest Worksheet		
Interest	I	
Principal	p	
Rate	r	
Time	t	

I	$=$	p	$\times$	r	$\times$	t

Compound Interest

Compound Interest Worksheet		
Balance	B	
Principal	p	
Rate	r	
Time	t	

B	$=$	p	$\times$	(1	$+$	r)	t (as an exponent)